AF506173

R. Kallenborn · H. Hühnerfuss

Chiral Environmental Pollutants

Springer
Berlin
Heidelberg
New York
Barcelona
Hong Kong
London
Milano
Paris
Singapore
Tokyo

R. Kallenborn · H. Hühnerfuss

Chiral Environmental Pollutants

Trace Analysis and Ecotoxicology

Springer

Roland Kallenborn
Norwegian Institute for Air Research (NILU),
Polar Environmental Centre
9296 Tromsø, Norway
e-mail: Roland.Kallenborn@nilu.no

Heinrich Hühnerfuss
University of Hamburg
Institute of Organic Chemistry
Martin-Luther-King-Platz 6
20146 Hamburg, Germany
e-mail: huehnerfuss@chemie.uni-hamburg.de

ISBN 3-540-66423-8 Springer-Verlag Berlin Heidelberg New York

Library of Congress Cataloging-in-Publication Data
Kallenborn, R. (Roland), 1960-
  Chiral environmental pollutants : trace analysis and ecotoxicology /
R. Kallenborn, H. Huehnerfuss.
      p. cm.
Includes bibliographical references and index.
  ISBN 3540664238 (alk. paper)
  1. Environmental toxicology. 2. Enantiomers--Toxicology. I.
Huehnerfuss, H. (Heinrich), 1944- II. Title.
  RA1226 .K35 2001
  615-9'5--dc21

Springer-Verlag Berlin Heidelberg New York
a member of BertelsmannSpringer Science+Business Media GmbH

© Springer-Verlag Berlin Heidelberg 2001
Printed in Germany

Coverdesign: design & production, Heidelberg
Typesetting: MEDIO, Berlin

Printed on acid-free paper   SPIN: 10694330   61/3020 ra – 5 4 3 2 1 0

Preface

In 1848, Louis Pasteur [1] was the first to report the separation of two different types of sodium ammonium tartrate, which he assumed to be related to each other like two mirror images. A few years later this phenomenon was denominated *chirality* (from the Greek word χειρ (= cheir) which means hand) by Lord Kelvin who used the following definition: "*I call any geometrical figure or any groups of points chiral and say it has chirality if its image in a plane mirror, ideally realised cannot be brought to coincide with itself.*" (Lord Kelvin, 1883, according to [2]). Subsequently, a large number of compounds were observed to fulfil these requirements. In biological systems this phenomenon of asymmetry had already been known for quite some time. Different snail species, for example, produce mirror-image forms, fossils like ammonites exhibit chiral shapes, and, last but not least, the human body, including feet, hands and ears, can be divided into two parts which can be regarded as two non-superimposable mirror images.

After making some breath-taking scientific findings, two young Chinese-American physicists, Tsung Dao Lee and Chen Ning Yang, were awarded the Nobel prize in 1957 when they proved that the parity of weak interactions is not preserved [3]; asymmetric approaches were also applied to elementary and quantum physical processes under certain conditions (see Sect. 5.2.1). The simple principle that may be inferred from symmetry considerations, according to which two structures can be identical but not superimposable, is nowadays a basic research objective in at least four main scientific disciplines: biology, chemistry, physics and mathematics.

In the last decade, new trace analytical methods have been developed to separate and detect the enantiomers of persistent organic pollutants in a large number of environmental samples at different trophic levels. In 1991, Faller et al. [4] and Kallenborn et al. [5] were the first to publish the successful separation of α-HCH in seawater and marine biota samples, respectively. These early publications already showed the potential of this relatively simple and robust enantioselective gas chromatographic method to play a major role in assessing enzymatic transformation processes of anthropogenic and natural organic pollutants in the environment.

However, to date, no efforts have been made to give a comprehensive survey of the state-of-the-art of enantioselective trace analysis and to reflect the future of this promising new field of analytical research. The present monograph intends to fill this obvious gap and aims to deliver a comprehensive tool for the benefit of experienced trace analysts as well as of interested students and non-chemists.

International research on chirality of xenobiotics today mainly focuses on the two following principal questions:

1 Why does chirality play such an outstanding role when biochemical transformation and accumulation processes of chiral organic pollutants in organisms, from bacteria up to human beings, are being discussed?

2 How can chirality as a basic selector for enzymatic processes be documented and proven for organic chemicals which are transformed and/or accumulated in ultra-trace concentrations?

These two questions will form the central theme of the present survey. We will mainly focus on the trace analytical and ecotoxicological aspects of molecular asymmetry. Since almost the entire research work is centrally related to chiral organic xenobiotics, we will largely confine ourselves to organic compounds, although in general parts of this book theoretical aspects of chiral inorganic substances will also be mentioned. (References used in the Preface can be found at the end of Chap. 1.)

Heinrich Hühnerfuss
Roland Kallenborn   Hamburg and Tromsø, October 2000

Acknowledgements

The authors wish to express their gratitude to all colleagues who contributed to the present endeavour with knowledge, ideas, information and advice. In particular, the help of the following colleagues is gratefully acknowledged: S. Allenmark, R.J. Baczuk, Å. Bergman, T.H. Bidleman, E. Dybing, W. Engewald, J. Gal, E. S. Heimstad, J.-E. Haugen, B. Koppenhoefer, W.A. König, C. Larssen, V. Meyer, A. Mosandl, M. Oehme, H. Parlar, G.G. Rimkus, M. Schlabach, V. Schurig, K. Stine, S. Tanabe, W. Vetter, and J. Whatley.

Several enthusiastic co-workers, who pioneered the enantioselective analysis of chiral environmental pollutants and chiral model substances, should also be mentioned: B. Bethan, S. Biselli, T. Ellerichmann, J. Faller, R. Gatermann, A. Gericke, F. Hoffmann, P. Ludwig, K. Möller, N. Peters, and B. Pfaffenberger.

Thanks are also extended to E.I. Hanssen, Ø. Hov, O.A. Braathen, and I.C. Burkow for their help and support during the realisation and the paper work involved in our book project. The idealistic and partial financial support of the Norwegian Institute for Air Research (NILU) is very much acknowledged.

During the last phase of the publication procedure, H. Dannhauer kept us "alive" by expending continuous encouragement as well as constructive criticism, and she participated in developing a "usable" index. This invaluable support is gratefully acknowledged.

In addition, we appreciate the opportunity to publish this monograph with Springer-Verlag and thank the editors, especially senior editor P. Enders, for the administrative support during the realisation of this book.

Without the extreme patience and the support of our families, this book would not have been accomplished in its present form.

Contents

Abbreviations

∞	infinity
α	chromatographic separation factor
α_1-AGP	α_1-acid glycoproteins
β-NF	β-Naphthoflavon
β-PMCD	permethylated β-cyclodextrin
μm	micrometer
σ	standard deviation
τ	half-life time
AChE	acetylcholinesterase
A-CEKC	affinity capillary electrokinetic chromatography
ARDRA	amplified ribosomal DNA restriction analysis
AES-NCP	Arctic Environmental Strategy's Northern Contaminant Programme
AHH	aryl hydrocarbon hydroxylase
AHDI	Phantolide®, 1-(2,3-dihydro-1,1,2,3,3,6-hexamethyl-1H-inden-5-yl)ethanone
AHTN	Tonalide®, 1-(5,6,7,8-tetrahydro-3,5,5,6,8,8-hexamethyl-2-naphthalenyl)ethanone
AIDS	acquired immuno deficiency syndrome
AMAP	Arctic Monitoring and Assessment Programme
APCI	atmospheric pressure chemical ionisation
ATII	Traseolide®, 1-[2,3-dihydro-1,1,2,6-tetramethyl-3-(1-methylethyl)-1H-inden-5-yl]ethanone
BBB	blood-brain barrier
BC	bromocyclen
BGB-172	chiral stationary phase for capillary GC (Trademark)
BPDM	benzphetamine N-demethylase
BMF	biomagnification factor
BSA	bovine serum albumin
C-18	octadecyl reversed stationary phase for liquid chromatography
Carbowax 20M	component of stationary phase for cGC
CAS	Chemical Abstract System
CC	*cis*-chlordane
CCE	chiral capillary electrophoresis

CD	cyclodextrin
CE	capillary electrophoresis
CEC	capillary electrochromatography
CEKC	capillary electrokinetic chromatography
cGC	capillary gas chromatography
CGE	capillary gel electrophoresis
CGTase	cyclodextrin-glucanosyltransferase
Chiralcel	stationary phases for liquid chromatography based on derivatised cellulose
Chiraldex G-PT	permethyltrifluoroacetoxypropyl-γ-cyclodextrin
Chirasil-Dex	chiral polysiloxanes containing chemically bonded permethylated β-cyclodextrin
Chirasil-METAL	metal(II)-bis[3-heptafluorobutanoylcamphorate]
Chirasil-Val®	stationary phase (copolymerisation of dimethylsiloxane with (2-carboxypropyl)methylsiloxane, coupling of the carboxy group with the amino group of L-valine-*tert*-butylamide)
CIEF	capillary isoelectric focusing
CIP	Cahn-Ingold-Prelog rules for the R,S nomenclature of chiral chemicals
CITP	capillary isotachophoresis
Cl_xBPE	chlorinated bis(propyl) ethers
CPDA	US Chemical Producers and Distributors Association
CPL	circularly polarised light
CPMP	Committee for Proprietary Medicinal Production
CSP	chiral stationary phase
CT	charge transfer
CTA	microcrystalline cellulose triacetate
CTT	technical Toxaphene: polychlorinated camphenes, bornanes and bornenes: Toxaphene, Melipax, Strobane
CYP	cytochrome P-450 enzymes, mixed functional oxidases
CZE	capillary zone electrophoresis
2,4-D	2,4-dichlorophenoxyethanoic acid
2,4-DCP	2,4-dichlorophenol
DCPP	2-(2,4-dichlorophenoxy)propanoic acid
o,p'-DDT	1,1,1-trichloro-2-(2-chlorophenyl)-2-(4-chlorophenyl)ethane
p,p'-DDT	1,1,1-trichloro-2,2-bis(4-chlorophenyl)ethane
o,p'-DDE	1,1-dichloro-2-(2-chlorophenyl)-2-(4-chlorophenyl)ethene
p,p'-DDE	1,1-dichloro-2,2-bis(4-chlorophenyl)ethene
o,p'-DDD	1,1-dichloro-2-(2-chlorophenyl)-2-(4-chlorophenyl)ethane
p,p'-DDD	1,1-dichloro-2,2-bis(4-chlorophenyl)ethane
o,p'-DDA	2-(2-chlorophenyl)-2-(4-chlorophenyl)ethanoic acid
p,p'-DDA	2,2-bis(4-chlorophenyl)ethanoic acid
dichlorprop	see DCPP
DMSO	dimethyl sulfoxide
DNA	desoxyribonucleic acid

DQF-COSY	double quantum filtered–correlated spectroscopy
EC	European Community
ECD	electron capture detector
ECNI-MS	electron capture negative ion mass spectrometry
ee	enantiomeric excess
EI	electron impact mode in mass spectrometry
EKC	electrokinetic chromatography
ELSD	evaporative light scattering detection
EOM	extractable organic material
US-EPA	US Environmental Protection Agency
ER	enantiomeric ratio
EROD	ethoxyresorufin-O-deethylase
ESI-MS	electrospray ionisation mass spectrometry
FAD	flavin adenine dinucleotide
FID	flame ionisation detector
FDA	US Food and Drug Administration
G	Gibbs free energy
GABA	γ-aminobutanoic acid
GFF	glass fibre filter
ΔH	heat of formation/enthalpy change
HCH	hexachlorocyclohexane isomers, a mixture with the main constituents: α-, β-, γ-, δ-, ε-HCH
HETP	height equivalent to the theoretical plate
HEPTA	heptachlor
HEPX	heptachlor *exo*epoxide
HHCB	Galaxolide®, 1,3,4,6,7,8-hexahydro-4,6,6,7,8,8-hexamethylcyclopenta[g]-2-benzopyran
HIV	human immunodeficiency virus
HPLC	high performance liquid chromatography
HPTLC	high performance thin layer chromatography
HRGC	high resolution gas chromatography
IADN	integrated atmospheric deposition network
IND	investigational exemption for a new drug
IRRAS	infrared reflection-absorption spectroscopy
IUPAC	International Union of Pure and Applied Chemistry
k'	capacity factor: $(t_R-t_0)/t_0$
K_{OW}	n-octanol/water partitioning coefficient
kV	kilovolt
LAH	lithium aluminum hydride
LC	liquid chromatography
LD	lethal dose, e.g., LD_{50}=lethal dose for 50% of the respective population
LEC	ligand exchange chromatography
Lipodex E	1:1 mixture of octakis(3-O-butyryl-2,6-di-O-n-pentyl)-γ-cyclodextrin and OV-1701

LOD	limit of determination
LSD	lysergic acid *N,N*-diethylamide
m	metre
M	molar
M1-At	1-hydroxy*iso*propylatrazine
MC 4	chlordane congener
MC 5	chlordane congener
MC 6	chlordane congener
MC 7	chlordane congener
MCPP	2-(4-chloro-2-methyl-phenoxy)propanoic acid
MDGC	multidimensional gas chromatography
MeSO$_2$-PCBs	methylsulfonyl PCBs
M-CEKC	micellar capillary electrokinetic chromatography
MFO	mixed functional oxidases
MHz	mega Hertz
MK	musk ketone
MPB	methylphenobarbitone
MS	mass spectrometry
MX	musk xylene
NAD	nicotineamide adenosine dinucleotide
NDA	new drug application
NICI	negative ion chemical ionisation
nL	nano-liter
NADPH	nicotinamide adenine dinucleotide phosphate, reduced
NIST	National Institute for Standards and Technology
NMR	nuclear magnetic resonance
NOE	nuclear Overhauser effect
NSAID	nonsteroidal antiinflammatory drug
N-TFA	*N*-trifluoroacetate
ODS	octadecyl silica
OSPAR	Oslo-Paris Commission
OXY	oxychlordane
OV-225	stationary phase for cGC fused-silica capillary columns
OV-1701	stationary phase for cGC fused-silica capillary columns
PAH	polynuclear aromatic hydrocarbons
PARLAR	nomenclature for toxaphenes (see text)
PB-1	P-450 isozymes
PB-4	P-450 isozymes
PCB	polychlorinated biphenyls, a mixture of 209 possible congeners
PCCH	pentachlorocyclohexene
PNA	peptide nuclear acid
PSO86	stationary phase for cGC fused-silica columns
QSAR	quantitative structure activity relationships
rac	racemic

RIFS	reflectometric interference spectroscopy
rRNA	ribosomal RNA
RNA	ribonucleic acid
RP	reversed phase
ΔS	entropy change
SANAE	South African National Antarctic Expedition
SD	standard deviation
SDS	sodium dodecyl sulphate
SE-30	stationary phase for cGC fused-silica capillary columns
SE-54	stationary phase for cGC fused-silica capillary columns
Sephadex	cross-linked dextran
SFC	supercritical fluid chromatography
SFE	supercritical fluid extraction
Silar-10C	stationary phase for cGC fused-silica capillary columns
SIM	selected ion monitoring
SIR	selected ion recording
SPMD	semipermeable membrane devices
SRM	standard reference materials
TAPA	tetranitro-9-fluorenylideneaminooxypropanoic acid
2,3,7,8-TCDD	2,3,7,8-tetrachloro-p-dibenzodioxin
TBDMS-CD	heptakis(6-O-*tert*-butyldimethylsilyl-2,3-di-O-methyl)-β-cyclodextrin
TC	*trans*-chlordane
TEF	toxic equivalency factor for dioxin-like PCBs and chlorinated dibenzo-p-dioxins
TEG-CD	octakis(2,3,6-tri-O-ethyl)-γ-cyclodextrin
TID	thermionic detector
TLC	thin layer chromatography
TSMR	thickness shear-mode resonance
URO	uroporphyrin
UV	ultraviolet light
U 81	chlordane congener
U 82	chlordane congener
ΔV	volume change
WCOT	wide coated open tubular columns
XE-60	stationary phase for cGC fused-silica capillary columns
XTI-5	stationary phase for cGC fused-silica capillary columns

1 Introduction

1.1
Links Between Chirality and Life Processes

In the old kingdoms of upper and lower Egypt, in many examples of burial chambers mural paintings depicting significant events, inscribed stelae or tablets have been excavated that show an astonishing proximity to our modern approach to chirality. In Fig. 1.1, an example is given of a stele that exhibits hieroglyphs carved out of stone such that they are mirrored on both sides of the so-

Fig. 1.1. Hieroglyphs carved out of stone on a stele outside the Pergamon museum, Berlin, Germany; translation: nsw.t biti ≡ king of upper and lower Egypt; the ankh cross in the centre is the symbol for life (courtesy Erika and Katja Hühnerfuss)

called "ankh cross". This ankh cross plays an important role in Egyptian mythology – it is the symbol for *life*. The pharaohs firmly believed in life after death, and, accordingly, they wanted to link the symbol for life with their coffins as well as with the mural paintings and stelae of the burial chamber. In addition, the hieroglyphs closely related to the ankh cross appear twice, i.e., in the form of their two mirror images. This ancient interpretation of chirality/life/death is well in line with the modern assumption that life processes, for example, enzymatic transformation processes, are highly enantioselective, but that dead organic matter may also be transformed and in part mineralised by very enantioselective processes.

Detailed evidence for the influence of symmetry and asymmetry on life processes and the obvious consequences for our daily life can be found in the realm of plants and animals, where numerous examples for symmetric and asymmetric macroscopic structures can be observed. For example, the occurrence of a horizontal radial symmetric shape for plants may be attributed to the influence of the sun light, which circles round the stem in the course of a day. Branches and leaves tend to grow towards the light, whereas roots are not at all attracted by light but rather follow the gravitational force. Wherever the symmetric lateral plane is placed, it will divide the object into (more or less) equal parts. On the other hand, a special case of radial symmetry, the so-called bilateral symmetry, is also often encountered in the realm of plants and animals. These kinds of objects only possess one vertical or horizontal internal plane of symmetry, sometimes referred to as "mirror plane". A symmetry of this kind can be defined for the human body. The symmetric plane divides the human body vertically into two more or less even parts with one ear, one eye, one nostril, one arm and one leg at each side. This type of symmetry is closely connected with the evolutionary development of the organism. In general, animals which are sessile and do not move are assumed to show a classical bilateral symmetry, while organisms which have developed locomotive properties are preferentially characterised by a radial symmetry. However, many exceptions to this rule can be found in natural systems, and, in addition, spherical, conical, cylindrical symmetries are also encountered.

On the other hand, macroscopic asymmetric structures possessing no plane of symmetry are also quite general in nature. Classic examples for asymmetric objects are helical structures, which are found for climbing and twining plants. For example, "honeysuckle (*Linnea borealis* L.)" always twines in a left-handed helix, whereas "morning glory (*Convulvulus sabatius* L.)" twines in a right-handed helix. Helical structures are also known for animals. Most prominent examples can be found for marine and freshwater molluscs (snails and shellfish). Normally, only one type of "handedness" is common for one species, but both right- and left-handed helical structures are known and found for all these asymmetric species. Several fossil molluscs demonstrate the evolutionary potential of asymmetric structures, e.g., ammonites and the *Nautilus pompilius* (L.) shell are the most prominent examples of asymmetric natural shapes in prehistoric time, where links have been revealed between ammonites and the present *cephalopodae*, while *Nautilus* represents one of those rare examples that have survived and almost exactly exhibit their former shape. Certain asymmetric helical structures can also be found for humans, e.g., the umbilical cord of the new-born child shows a clear helical structure. Cords have been found to be wound left- and right-handed. The reason why a certain direction is preferred is still under discussion [6].

Another type of non-helical asymmetry is represented by a passerine bird species which can be found both in North America and Middle Europe, a bird species called the crossbill (*Loxia curvirostra* L.) [3]. The bird's upper and lower beaks cross over each other, where two ways are possible, to the left- or the right-hand side of each other. These two ways represent two mirror-image-like procedures. Interestingly enough, the upper bill of the American population crosses to the left-hand side of the bird, whereas the opposite can be observed for the European birds [3].

An unusual type of asymmetric shape occurs in the marine flatfish family (*pleuronectiformes*); e.g., halibut (*Hippoglossus hippoglossus* L.), sole (*Glyptocephalus cynoglossus* L.) and flounder (*Pleuronectes platessa* L.) belong to this group of marine fish species. In an early life stage, one eye slowly migrates to the other side of the organism. As an adult animal, the flatfish lies on the ground with one side and both eyes upwards, hunting small crustaceans and fish. The side which is preferentially directed upwards differs from one species to the other; however, this side is characteristic of one species. "Left- and right-sided" individuals are known for several species (e.g., soles).

In general, it can be assumed for plants and animals that a radial or bilateral symmetry is coupled with small asymmetric structures [3]. During the evolution of present plants and animals, bilateral structures, in particular, have proven to be important for survival strategies. Nevertheless, there cannot be any doubt that for certain purposes asymmetric structures have shown to be better adapted than bilateral symmetric shapes.

The first to discover the principle of asymmetry from a chemical point of view was Louis Pasteur, who succeeded in separating two types of sodium ammonium tartrate crystals in 1848. He showed that separate solutions of these two types of crystals are able to rotate the plane of linearly polarised light in different directions, to the right- and left-hand side, respectively [1]. Depending on this characteristic direction the two isomers were called 'left-handed' and 'right-handed'. The principle properties of chirality were found. In a subsequent experiment Pasteur discovered that mould degrades only one type of molecule (left- or right-handed), whereas the other remains intact. This was without any doubt the first step in the direction of modern biochemistry [2]. A few years later, in 1883, this phenomenon was named chirality by Lord Kelvin (after the Greek word for hand). At that time, the majority of scientists did not consider these results to be very important for biological processes. However, a vivid discussion was initiated, where chemists and biologists discussed the implication of chirality for natural processes. These discussions lasted a relatively long time (until the early years of the 20th century) and were one of those typical examples in the history of science, where unconventional researchers, leaving the traditional ways of scientific thinking, had to fight against ignorance raised by the established scientific community and had to struggle in order to break down the wall of traditions.

1.2
General Principles of Chirality

Nowadays, the principles of chirality are included in basic lectures of organic chemistry. Therefore, the authors of the present monograph do not intend to duplicate these basic lectures. On the other hand, scientists of other disciplines may not be that familiar with the terminology and basic rules, which are assumed to be crucial for a comprehensive understanding of the molecular structures that determine chiral molecules and enantioselective processes, as discussed here. Therefore, we have decided to set a uniform "stage" for readers of all disciplines by giving a brief survey of some crucial aspects ruling chiral molecules and enantioselective processes. In order to meet the requirements of the present work as closely as possible, examples for the different types of chirality will be represented by chiral environmental pollutants discussed herein.

Chirality is a very important field of stereoisomerism according to the following definitions: stereoisomers are compounds made up of the same atoms, bonded by the same sequence of bonds, but possessing different three-dimensional structures which are not interchangeable. These three-dimensional structures are called configurations. In the same way as many things around us, such as our hands and pairs of shoes, are not identical, but a mirror image of one another, non-identical stereoisomers exist in which the only distinction between them is that one is the mirror image of the other. However, these mirror images are not superimposable. A simple example of this type of stereoisomerism is represented by the herbicide dichlorprop (or DCPP), i.e., 2-(2,4-dichlorophenoxy)propanoic acid (Fig. 1.2), which can exist in two spatial configurations that correspond to reflections of each other. These stereoisomers are specifically called enantiomers. A 1:1 mixture of both enantiomers forms a racemate or ra-

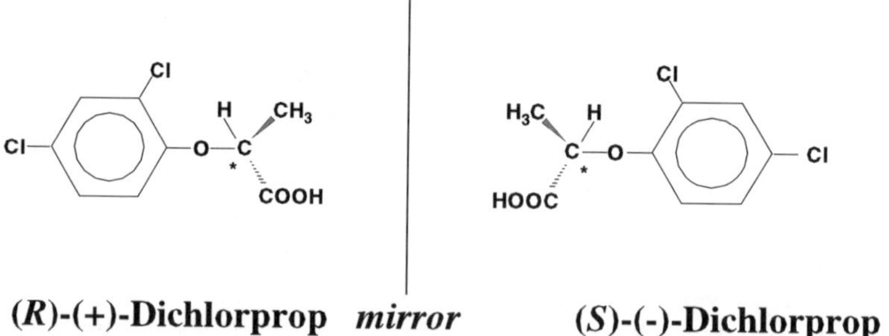

Fig. 1.2. Non-superimposable mirror images, so-called (R)- and (S)-enantiomers, of the herbicide dichlorprop [or DCPP, i.e., 2-(2,4-dichlorophenoxy)propanoic acid]

cemic mixture. In addition to these general definitions, some rules have to be summarised that allow a discrimination between chiral and achiral molecules.

1.2.1
Chiral Environmental Pollutants with a Stereogenic Centre

The most common origin of chirality in molecules, and the one originally recognised by van't Hoff and Le Bel (1874), is the presence of one or more atoms, for example tetraedral carbon atoms, each of which forms noncoplanar bonds to *four different atoms or groups*. The atom that carries the four different substituents is called the *asymmetric* or *stereogenic centre*. This is the case for the example given in Fig. 1.2, where the stereogenic centre bonds to a hydrogen atom, a methyl, a carboxyl and a 2,4-dichlorophenoxy group. In evaluating a chemical structure for chirality, a carbon carrying four different attached groups may give one indication of the presence of a chiral compound.

Many important natural compounds, such as amino acids and carbohydrates, possess one or more stereogenic centres linked with four different atoms or groups. However, several other environmental pollutants discussed in the present work exhibit this stereochemical characteristic, which can also be easily recognised by inexperienced readers. A list of these environmental pollutants can be found in Table 1.1.

Furthermore, it is important to note that one of the four different atoms and groups, respectively, carried by the stereogenic centre may be substituted by a free pair of electrons. This type of chiral compound can be found in the homologous series of tertiary amines or sulfoxides. A parameter that limits the possibility of isolating and analysing enantiomers by enantioselective chromatographic approaches is the inversion barrier. Fast inversion between the two enantiomers would render it impossible to separate them chromatographically. Examples of this type of chirality will be discussed in Sections 4.1 (naloxone; one stereogenic nitrogen centre) and 4.6, where the results reported by Waxman et al. [7] are summarised. These authors studied the enantioselective sulfoxidation of 4-tolylethyl sulfide applying two cytochrome P-450 isoenzymes purified from phenobarbital-induced rat liver, both of which generated 4-tolylethyl sulfoxide (Fig. 1.3), exhibiting predominantly the (*S*)-(–)-configuration. An example of a stereogenic phosphorus centre is represented by ruelene, i.e., (*R,S*)-4-*tert*-butyl-2-chlorophenylmethyl-*N*-methyl phosphoramidate (Sect. 3.5.2).

1.2.2
Environmental Pollutants with Axial Chirality

Certain compounds that do not contain asymmetric atoms may nevertheless be chiral if they contain a structure represented, for example, by a sub-group of polychlorinated biphenyls (PCBs; Fig. 1.4) and their metabolites. Biphenyls containing four large groups in *ortho* positions cannot freely rotate about the central single bond because of steric hindrance. In such compounds the two ring sys-

Table 1.1. Chiral environmental pollutants discussed in the present work with one or more stereogenic centres linked with four different substituents; examples of *central asymmetry*

Chiral environmental pollutant	Section(s)
Dichlorprop [DCPP; i.e., (*R*,*S*)-2-(2,4-dichlorophenoxy)propanoic acid]	3.1; 3.5
Methyl dichlorprop [i.e., methyl-(*R*,*S*)-2-(2,4-dichlorophenoxy)propanoate]	3.5
MCPP [i.e., 2-(4-chloro-2-methyl-phenoxy)propanoic acid]	3.1
Ibuprofen, (i.e., 2-[4-(2-methyl)propyl]propanoic acid) and metabolites	3.3
4-Tolylethyl sulfoxide (stereogenic sulfur centre)	4.6
Ruelene, [i.e., (*R*,*S*)-4-*tert*-butyl-2-chlorophenylmethyl-*N*-methyl phosphoramidate] (stereogenic phosphorus centre)	3.5
Chlorinated bis(propyl) ethers	3.1
o,p'-DDT	2.1; 2.3; 2.4; 3.1; 3.2
o,p'-DDD	2.3; 2.4; 3.1; 3.2
Chlordanes (*cis*-, *trans*-, other congeners)	2.1; 3.2; 3.5; 3.6
Oxychlordane	3.2; 3.5;
Chlordene and metabolites	4.6
Photochlordene	3.4
Heptachlor	3.2; 4.6
Heptachlor *exo*-epoxide	3.2; 3.5; 4.6
Photo-heptachlorepoxide	3.4
Bromocyclen	3.1
Toxaphenes	2.1; 3.2; 3.5
HHCB (1,3,4,6,7,8-hexahydro-4,6,6,7,8,8-hexamethylcyclopenta[*g*]-2-benzopyran; Galaxolide®; two stereogenic centres)	3.3
Galaxolidone (1,3,4,6,7,8-hexahydro-4,6,6,7,8,8-hexamethylcyclopenta[*g*]-2-benzopyran-1-one metabolite of Galaxolide; two stereogenic centres)	3.3
AHTN [1-(5,6,7,8-tetrahydro-3,5,5,6,8,8-hexamethyl-2-naphthalenyl)ethanone; Tonalide®]	3.3
ATII (1-[2,3-dihydro-1,1,2,6-tetramethyl-3-(1-methyl-ethyl)-1*H*-inden-5-yl]ethanone; Traseolide®; two stereogenic centres)	3.3
AHDI [1-(2,3-dihydro-1,1,2,3,3,6-hexamethyl-1*H*-inden-5-yl)ethanone; Phantolide®]	3.3
Thalidomide (Contergan®)	4.1
Naloxone (three stereogenic carbon centres, one nitrogen centre)	4.1
Deltamethrin (three stereogenic carbon centres)	4.2
Tetrodotoxin	4.5
Saxitoxin	4.5
Anatoxin-*a* and homoanatoxin-*a*	4.5

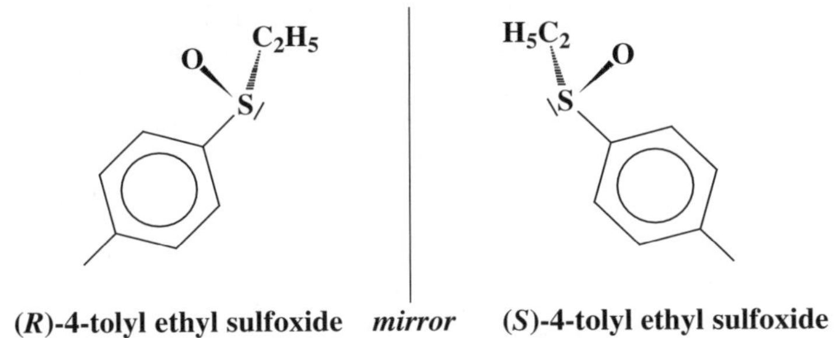

(R)-4-tolyl ethyl sulfoxide   *mirror*   **(S)-4-tolyl ethyl sulfoxide**

Fig. 1.3. Sulfoxidation products of 4-tolylethylsulfide, (R)-4-tolylethyl sulfoxide (left-hand side) and (S)-4-tolylethyl sulfoxide (right-hand side); for details, see Sect. 4.6

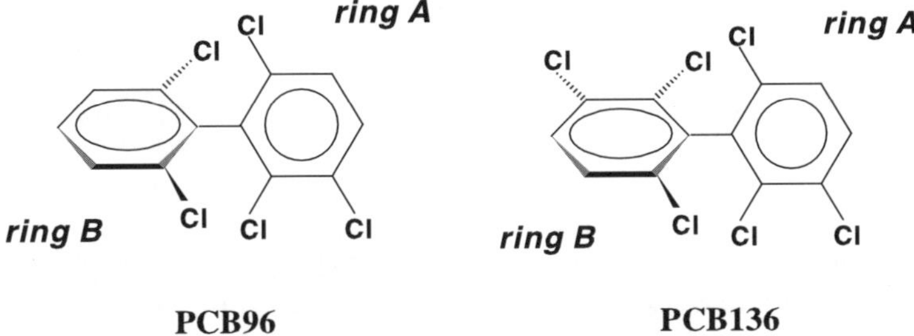

PCB96   **PCB136**

Fig. 1.4. The PCB congener on the left-hand side possesses a *plane of symmetry* perpendicular to ring B, which is assumed to be oriented perpendicular to ring A, and hence PCB96 is achiral. By contrast, PCB136 possesses no plane of symmetry and thus is a chiral PCB congener

tems are oriented in perpendicular planes or at least in planes between which an angle of >0 to 90° exists. If either or both rings are symmetrical and perpendicular to each other, the molecule has a so-called *plane of symmetry*. For example, in PCB96 (Fig. 1.4) ring B is symmetrical. A plane drawn perpendicular to ring B such that the part on one side of the plane is the exact reflection of the part on the other side contains all the atoms and groups in ring A, i.e., this plane would cut the molecule into two equal parts. A mirror image of PCB96 reflected parallel to this *plane of symmetry* would not meet the requirement of being non-superimposable with the original molecule, and hence this compound is achiral. On the other hand, if an asymmetric substitution of both ring systems in a PCB is encountered, as represented by PCB136 in Fig. 1.4, no plane of symmetry exists and, as a consequence, this PCB congener is chiral.

It is not always necessary for four large *ortho* groups to be present in order to inhibit free rotation. Compounds with three and even two groups, if large enough, can exhibit hindered rotation and, if suitably substituted, can be resolved into the two enantiomeric forms. Basically 78 PCB congeners out of 209 are chiral; however, only 19 turned out to be stable against racemisation at room temperature (Table 1.2). The rotational barrier of the latter group of PCBs at-

Table 1.2. IUPAC nomenclature and chlorine substitution pattern of atropisomeric PCB congeners that are stable against racemisation at room temperature

PCB congener	Chlorine substitution	PCB congener	Chlorine substitution
PCB 45	2,2',3,6	PCB 144	2,2',3,4,5',6
PCB 84	2,2',3,3',6	PCB 149	2,2',3,4',5',6
PCB 88	2,2',3,4,6	PCB 171	2,2',3,3',4,4',6
PCB 91	2,2',3,4',6	PCB 174	2,2',3,3',4,5,6'
PCB 95	2,2',3,5',6	PCB 175	2,2',3,3',4,5,6
PCB 131	2,2',3,3',4,6	PCB 176	2,2',3,3',4,6,6'
PCB 132	2,2',3,3',4,6'	PCB 183	2,2',3,4,4',5',6
PCB 135	2,2',3,3',5,6'	PCB 196	2,2',3,3',4,4',5,6
PCB 136	2,2',3,3',6,6'	PCB 197	2,2',3,3',4,4',6,6'
PCB 139	2,2',3,4,4',6		

tains values of at least 105 to 250 kJ/mol [8]. PCB136, shown in Fig. 1.4, belongs to this group. PCB stereoisomers that can be separated because rotation about single bonds is prevented or greatly slowed are called atropisomers. Very detailed investigations on the rotation barriers in relation to the substituents and the substitution pattern of biphenyl congeners were carried out by the research groups of Schurig et al. [9], König and co-workers [10, 11] and Harju and Haglund [8]. For details on this aspect the reader is referred to these four publications (see also Sect. 3.2.1.1). Furthermore, it should be noted that if either ring is symmetrical with respect to *ortho-* and *meta*-positions, the molecule also possesses a plane of symmetry if an additional group in the *para*-position is present, because such groups cannot cause lack of symmetry.

1.2.3
Asymmetry of Cyclic Environmental Pollutants

Sometimes it is difficult to recognise the asymmetry of certain cyclic environmental pollutants such as hexachlorocyclohexanes (HCH), because some additional rules must be applied in order to discriminate reliably between achiral and chiral stereoisomers. Although the ultimate criterion is, of course, non-superimposability on the mirror image, simple tests like the *presence of a plane of symmetry* (see previous section) may often be successfully applied. For instance, the plane of symmetry can be easily found in molecules of the two most prominent HCH isomers, γ-HCH and β-HCH, presumably also by scientists who are not so familiar with chiral compounds: if we adopt the nomenclature of carbohydrates, where the two possible chair conformations are denominated $^4C_1$ (conformation, in which the C4 is drawn upwards and the C1 downwards as

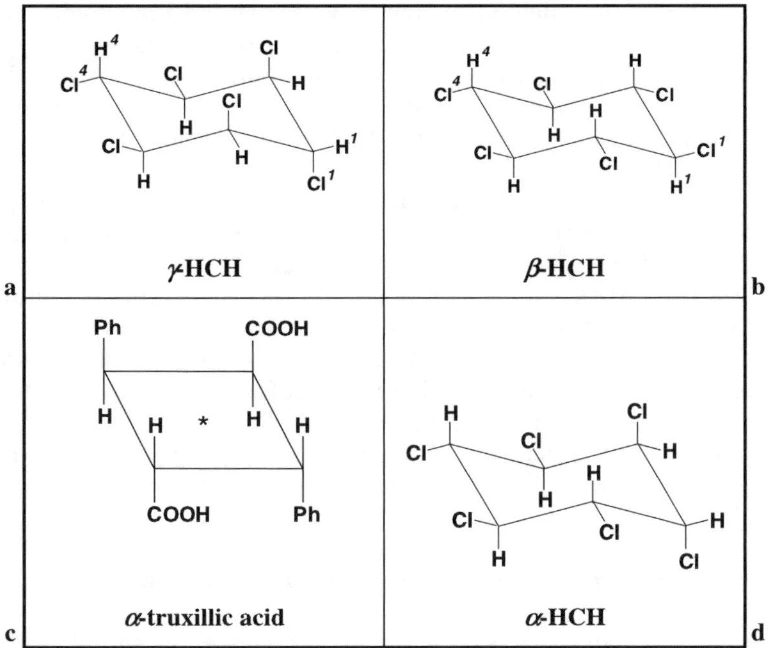

Fig. 1.5a–d. a γ-HCH possessing a *plane of symmetry* (H^1, Cl^1, C^1, H^4, Cl^4, C^4); **b** β-HCH exhibiting a *plane of symmetry* (H^1, Cl^1, C^1, H^4, Cl^4, C^4); **c** α-truxillic acid possessing a *centre of symmetry* (asterisk *); **d** α-1,2,3,4,5,6-hexachlorocyclohexane (α-HCH) exhibiting neither a *plane* nor a *centre* nor *an alternating axis of symmetry*

shown in Fig. 1.5) and $^1C_4$, we can define exactly the plane of symmetry, which in both cases can be laid through the H^1, Cl^1, C^1, H^4, Cl^4, C^4 atoms. Hence, these two HCH isomers, as well as five other HCH isomers where similar tests may be applied, are achiral.

On the other hand, this simple test may be misleading. Compounds possessing such a plane are always achiral, but a few cases are known in which compounds lack a plane of symmetry and are nevertheless achiral. Such substances may possess a centre of symmetry, for instance α-truxillic acid (asterisk in Fig. 1.5), a component of the cocaine alkaloids. A centre of symmetry is a point within an object such that a straight line drawn from any part or element of the object to the centre and extended an equal distance on the other side encounters an equal part or element. Furthermore, it has to be determined as to whether or not an alternating axis of symmetry can be found. An alternating axis of symmetry of the order n (also called an improper axis of rotation) is an axis such that when an object containing such an axis is rotated by 360/n about the axis and then reflection is effected across a plane at right angles to the axis, a new object is obtained that is indistinguishable from the original one. In general, a one-fold alternating axis is equivalent to a plane of symmetry and a two-fold alternative axis is identical with a centre of inversion. Therefore, the fundamental symmetry condition for optical activity is the absence of an improper axis. In the case of α-HCH, neither a plane nor a centre nor an alternating axis of symmetry is encountered, and, as a consequence, α-HCH is chiral, the only chiral of eight conceivable HCH isomers.

In lectures and monographs of basic organic chemistry (see for example [12, 13]), additional types of chirality are usually discussed. For completeness, we briefly mention these types without going into detail, because we do not necessarily need these aspects for the present purposes. These types of chirality include allene-type chirality, in which the central carbon atom is sp-bonded with even numbers of double bonds, both sides being asymmetrically substituted, helical asymmetry (represented by hexahelicene), and planar asymmetry (represented by substituted paracyclophanes). Basically, the same tests as those described above (plane or centre or alternating axis of symmetry) can be applied in order to find out whether or not such molecules are chiral.

1.2.4
Chiral Environmental Pollutants with Two or More Stereogenic Centres

When a molecule possesses two asymmetric centres, each centre has to be attributed its own configuration. The first centre may exhibit R- or S-configuration, and so may the second. Accordingly, a systematic variation of all possibilities at each stereogenic centre shows that two times two, i.e., four, stereoisomers can be formulated. Generalising, the maximum number of stereoisomers existing for a molecule with n stereogenic centres thus may be 2^n stereoisomers. In order to distinguish between these different possibilities, the R,S-nomenclature that is based upon the so-called *CIP rules* is strongly recommended, because these rules can be universally used also in those cases where the old D,L-nomenclature is ambiguous. It would be beyond the scope of the present monograph to develop the exact CIP rules here. For descriptions of the system and sets of sequence rules the reader should refer to basic organic monographs, e.g., [12, 13], or to the original publications by Cahn, Ingold and Prelog (= CIP) [14–16]. In the latter case, the more complicated assignments such as those for $(+)$-α-$1S,2R,3R,4S,5S,6S$-HCH will become accessible, though some time will be required to infer the exact nomenclature for such cyclic compounds.

Although four is the maximum number of isomers when the compound exhibits two chiral centres, this number may be reduced when the three groups on one stereogenic centre are the same as those on the other. In this case, one of the isomers has a plane of symmetry and hence is achiral, even though it exhibits two asymmetric centres. A typical example, tartaric acid, is shown in Fig. 1.6, for which only three isomers exist: a pair of enantiomers and an achiral so-called *meso*-form. Therefore, it is strongly recommended that one should exclude or verify the existence of a *meso*-form, if chiral compounds with two stereogenic centres are being investigated, where the four groups on one asymmetric centre

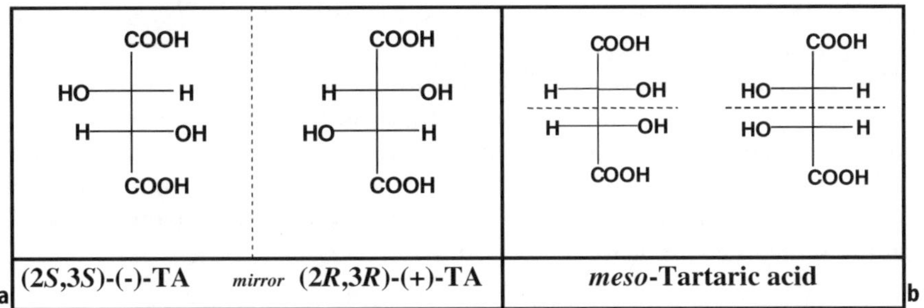

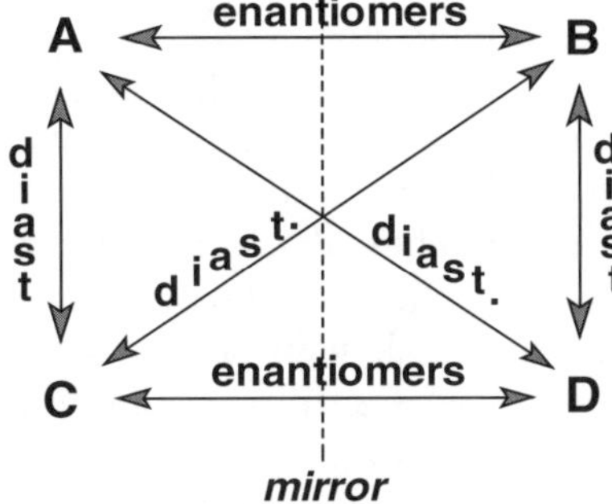

Fig. 1.6a,b. a the non-superimposable mirror images (= enantiomers) (2*S*,3*S*)-(–)- and (2*R*,3*R*)-(+)-tartaric acid; **b** the two *meso*-forms of tartaric acid which exhibit a *plane of symmetry* (- - - -)

Fig. 1.7. Relationship between the four stereoisomers A, B, C and D that can be formulated for chiral compounds with two stereogenic centres and thus $2^n=4$ stereoisomers. A/B and C/D are mirror images of each other and thus *enantiomers*, while all other relationships are designated by the term *diastereomers* (diast.)

are the same as those on the other chiral atom. This confirmation is necessary because this may have consequences for enantioselective chromatography, an aspect which will be discussed in more detail below.

In the case of a molecule with two stereogenic centres, four stereoisomers, A, B, C and D, can be formulated, as outlined above. Since a molecule can only possess one mirror image, each of the four stereoisomers can only be the reflection of one of the three remaining stereoisomers. As a consequence, two enantiomeric pairs exist, A/B and C/D. As schematically shown in Fig. 1.7, the pairs A/C, A/D, B/C and B/D cannot be mirror images of each other. These relationships are designated by the term diastereomers. It is important to note that diastereomers in general possess different physical properties. In contrast, enantiomers exhibit identical physical and chemical properties except in two important respects:

– They rotate the plane of linearly polarised light in opposite directions, though in equal amounts. The isomer that rotates the plane to the left (counterclockwise) is called the *levo-isomer* which is indicated with (–), while the one that

rotates the plane to the right (clockwise) is called the *dextro-isomer* and is designated (+).
- They react at different rates with other chiral compounds. These rates may be so close that the distinction is practically useless, or they may be so far apart that one enantiomer undergoes the reaction at a convenient rate, while the other does not react at all. This is the reason why many compounds are biologically active while their enantiomers are not. However, enantiomers react at the same rate with achiral compounds.

Furthermore, enantiomers may react at different rates with achiral molecules, if a chiral catalyst is present, they may show different solubilities in a chiral solvent, they may exhibit different indexes of refraction or absorption spectra when examined with circularly polarised light. In most cases these differences are too small to be useful and are often too small to be measured. In general, it can be concluded that enantiomers possess identical properties in a symmetrical environment, but their properties may differ in an unsymmetrical environment. This basic principle forms the background for enantioselective chromatography, where chiral stationary phases are used which form the asymmetric environment, as required. Thus, diastereomeric complexes between the respective enantiomers and the chiral stationary phase emerge that give rise to different retention times, hence allowing an enantioselective separation. By contrast, diastereomers which posses different physical properties can also be separated on achiral stationary phases.

The basic principles of chirality summarised above set the stage for an understanding of the different chromatographic characteristics of chiral and achiral environmental pollutants with one or more stereogenic centres. As the separation of diastereomers can be accomplished on achiral stationary phases, the maximum number of peaks to be expected will be identical with the number of diastereomers. The separation of enantiomers requires an asymmetric environment and thus chiral stationary phases. Then, the maximum number of peaks to be expected will be 2^n, provided that no *meso*-forms are present and accordingly reduce this number. An example of a chiral environmental pollutant that exhibits two asymmetric centres, and which can be successfully separated into its two diastereomeric pairs of enantiomers (i.e., four peaks) by enantioselective capillary gas chromatography, is represented by the musk compound HHCB (Fig. 1.8; for further details, see Sect. 3.3). Bis(2,3-dichloro-1-propyl) ether also possesses two asymmetric centres; however, in this case, a *meso*-form is encountered (Fig. 1.8; $2R,2'S \equiv 2S,2'R$). As a consequence, only three peaks have to be expected in the chromatograms obtained by enantioselective capillary gas chromatography. The assignment of the peak in the gas chromatogram to the *meso*-form was easy, because the peak height was approximately twice that of the enantiomers of the chiral stereoisomer (see Sect. 3.1.2).

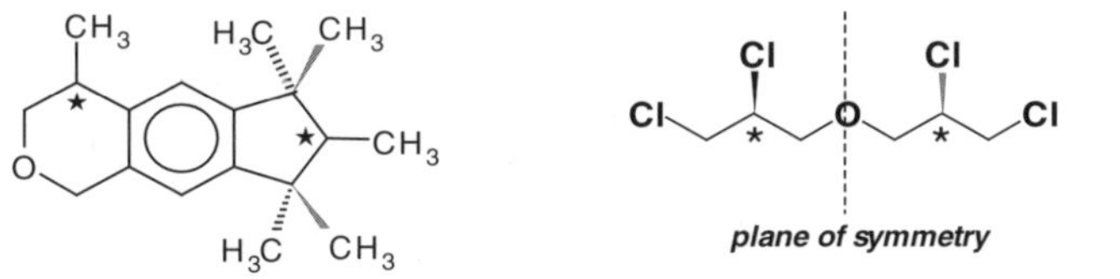

Fig. 1.8. The molecular structures of the musk compound HHCB (Galaxolide®) and bis(2,3-dichloro-1-propyl) ether which possess two diastereomeric centres (asterisks *)

References

1. Pasteur L (1848), Mémoire sur la relation qui peut exister entre la forme crystalline et la composition chimique, et sur la cause de la polarisation rotatoire. Comptes Rend Acad Sci 26:535–538
2. Jacques, J (1993) The molecule and its double. McGrawHill, p 128
3. Gardner, M (1990) The ambidextrous universe: mirror asymmetry and time-reversed worlds. Scribner, New York, p 293
4. Faller J, Hühnerfuss H, König WA, Krebber R, Ludwig P (1991) Do marine bacteria degrade α-hexachlorocyclohexane stereoselectively? Environ Sci Technol 25:676–678.
5. Kallenborn R, Hühnerfuss H, König WA (1991) Enantioselective metabolism of (±)-α-hexachlorocyclohexane in organs of the Eider Duck. Angew Chem 103:328–329; Angew Chem Int Ed Engl 30:320–321
6. Fletcher S (1993) Chirality in the umbilical cord. Br J Obstet Gynaecol Mar 100:234–236
7. Waxman DJ, Light DR, Walsh C (1982) Chiral sulfoxidations catalyzed by rat liver cytochromes P-450. Biochemistry 21:2499–2507
8. Harju MT, Haglund P (1999) Determination of rotational barriers of atropisomeric polychlorinated biphenyls. J Anal Chem 364:219–223
9. Schurig V, Glausch A, Fluck M (1995) On the enantiomerization barrier of atropisomeric 2,2',3,3',4,6'-hexachlorobiphenyl (PCB 132). Tetrahedron Asymm 6:2161–2164
10. Weseloh G, Wolf C, König WA (1996) New technique for the determination of interconversion processes based on capillary zone electrophoresis: studies with axially chiral biphenyls. Chirality 8:441–445
11. Wolf C, Hochmuth DH, König WA, Roussel C (1996) Influence of substituents on the rotational energy barrier of axially chiral biphenyls. Part 2. Liebigs Ann 1996:357–63
12. Walter W, Francke W (1998) Beyer-Walter, Lehrbuch der Organischen Chemie, 23rd edn. S Hirzel Verlag, Stuttgart, p 1176
13. March J (1985) Advanced organic chemistry. Reactions, mechanisms, and structure. Wiley, New York, p 1346
14. Cahn RS, Ingold CK, Prelog V (1956) The specification of asymmetric configuration in organic chemistry. Experientia 12:81–124
15. Cahn RS, Ingold CK, Prelog V (1966) Spezifikation der molekularen Chiralität. Angew Chem 78:413–447; Angew Chem Int Ed Engl 5:385–415
16. Prelog V, Helmchen G (1982) Grundlagen des CIP-Systems und Vorschläge für eine Revision. Angew Chem 94:614–631; Angew Chem Int Ed Engl 21:567–583

2 Enantioselective Chromatographic Methods for the Analysis of Chiral Environmental Pollutants

2.1
General Considerations

Enantioselective separations by chromatographic methods can be achieved using two different experimental approaches [1]. Firstly, diastereomeric derivatives are formed by reaction of a chiral compound with a chiral reagent. They may be separated on a stationary phase, which needs not necessarily be chiral. The advantage of this procedure is that a separation is possible with any chromatographic system as long as the required selectivity is available. The presence of at least one functional group as a site of reaction with an optically pure reagent is mandatory if the substrate is to be separated. One of the main drawbacks is the need to use optically pure reagents in order to avoid a systematic error. Another error may be introduced by the generation of energetically different (diastereomeric) transition states in the reaction of a mixture of enantiomers with a chiral reagent. Unless the reaction proceeds to completion, the different rates of these reactions may cause kinetic resolution and result in erroneous proportions of the product.

A second conceivable approach to stereochemical analysis consists of direct enantiomer separation by using chiral stationary phases. In this case, the separation is caused by the formation of diastereomeric, and thus energetically unequal, association complexes between the enantiomers to be separated and a chiral stationary phase. It is important to note that separation of the optical antipodes may occur as a result of *chiral recognition processes* (for details see, Sect. 2.2.3).

König pointed out in his monograph that even if the chiral stationary phase is not a pure enantiomer this would not affect the accuracy of the result; the separation factors α would merely be diminished [1]. On the other hand, variable separation properties between columns containing the same cyclodextrin derivative originating from different batches or manufacturers have caused problems [2–4]. The separation of polychlorinated pesticides may react very sensitively to small changes in the selectivity and polarity of the stationary phase. Differences in the functional group distribution of incompletely derivatised heptakis(2,3,6-

O-tert-butyldimethylsilyl)-*β*-cyclodextrin molecules may even lead to reversed elution orders of *cis*-chlordane [4] and *α*-HCH [3] enantiomers. A significant influence of phase purity and composition on the separation properties has also been reported for other cyclodextrin derivatives such as heptakis(6-*O-tert*-butyldimethylsilyl-2,3-di-*O*-methyl)-*β*-cyclodextrin (TBDMS-CD).

Substantial differences in the enantioselectivity for toxaphenes were observed between a semi-raw and a purified product [5]. Other commercially available cyclodextrin derivatives are also often mixtures of differently substituted products [6]. From these observations, Jaus and Oehme drew the following conclusions [2]: only cyclodextrin derivatives of known and reproducible composition should be used. This requires a control of purity and composition by methods such as high performance TLC, high temperature HRGC and HPLC combined with refractive index or evaporative light scattering detection (ELSD). The application of supercritical fluid chromatography (SFC) with ELSD has also been reported. In addition, compound identification has been carried out by fast bombardment mass spectrometry, electrospray ionisation mass spectrometry (ESI-MS) (also combined with HPLC) and capillary electrophoresis (CE) coupled to ESI-MS. For literature describing the methods summarised above, the reader should refer to [2].

During the systematic search for a substitute for TBDMS-CD, octakis(2,3,6-tri-*O*-ethyl)-*γ*-cyclodextrin (TEG-CD) was found to be an interesting alternative [2,7]. Incompletely derivatised batches especially showed a comparable enantioselectivity and thermal stability. In contrast to TBDMS-CD, no degradation over time was observed, and columns could be reproduced from the same batch over a period of months. The different TEG-CD batches were characterised by HPLC-ELSD or HPLC-MS by Jaus and Oehme [2], who also studied the influence of the composition on the enantioselectivity for the polychlorinated pesticides. As shown in Fig. 2.1, widely differing compositions of TEG-CD can be obtained depending on the degree of ethylation. The identity of the compounds was elucidated by LC/MS in the APCI(+) mode: Batch 1 contained nearly pure TEG-CD. Only a small shoulder of a product with one methyl group instead of ethyl was identified, representing about 5%. Batch 2A (Fig. 2.1) consisted of only 18% of TEG-CD which eluted last. The rest were under-ethylated products with one to four free hydroxy groups. The most abundant congeners had one to three ethyl groups less. In fraction 2B (Fig. 2.1), TEG-CD was hardly detectable. The main compounds were identified as products with two to five free hydroxy groups.

Jaus and Oehme expected that the three TEG-CD batches would show a different polarity and enantioselectivity when used as stationary phases in HRGC [2]. Two capillary columns of about 12 m length each were coated with material from each batch, in order to control the reproducibility of the manufacturing process. First, a Grob test was carried out resulting in separation numbers for the methyl esters between 23 and 27. The purpose of the Grob test is to determine the quality and/or monitor the performance and deterioration of a column during use. Column efficiency, activity and film thickness are easily evaluated using an appropriately designed test mixture [8, 9]. The various TEG-CD batch-

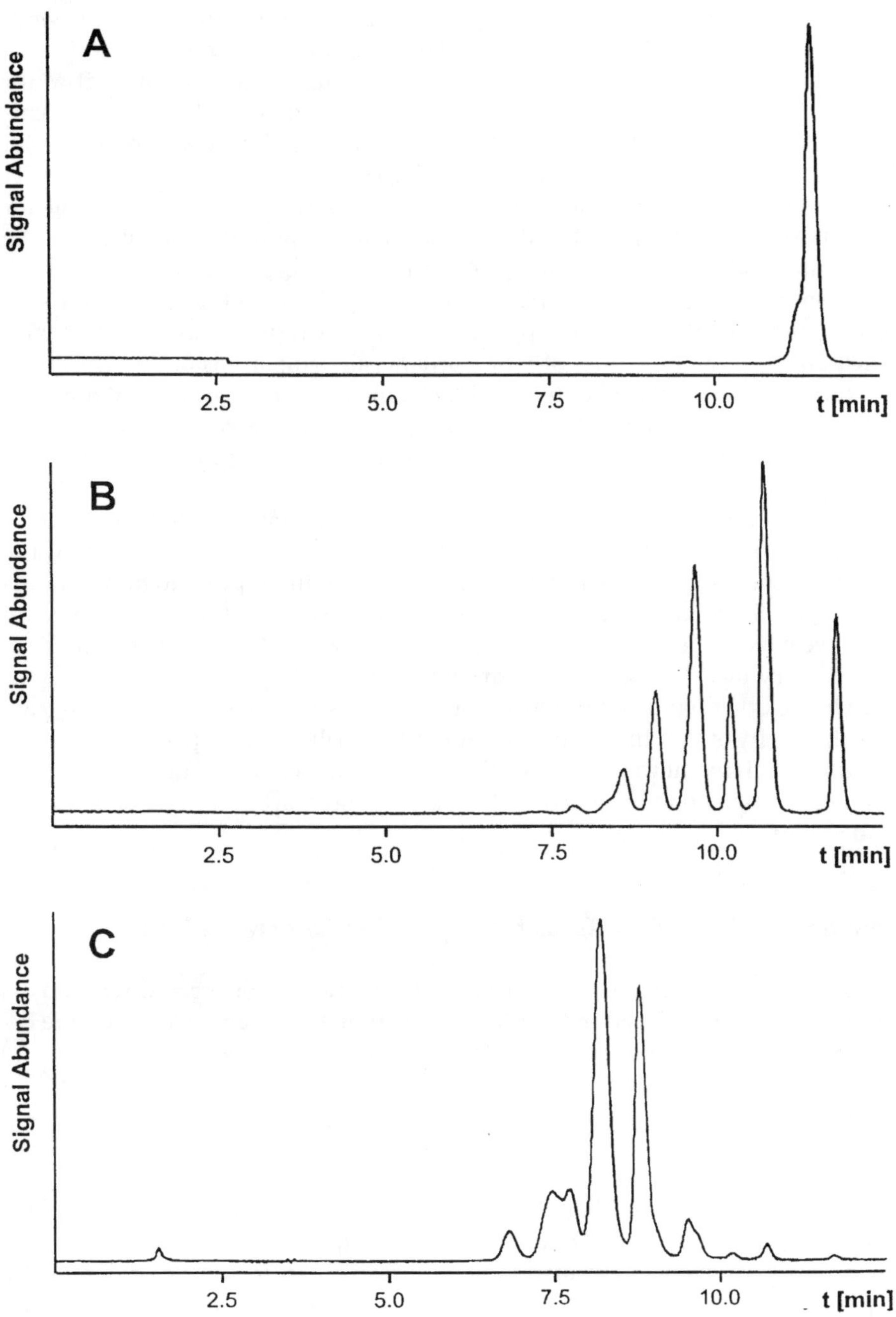

Fig. 2.1. HPLC-ELSD chromatograms of **A** TEG-CD batch 1, **B** TEG-CD batch 2A and **C** TEG-CD batch 2B [2]

es showed only small and insignificant differences in the results from the Grob test. Capillaries coated with the same batch material were nearly identical.

As the next step, Jaus and Oehme tested the enantioselective properties for polychlorinated pesticides using an appropriate test mixture, denominated "chirotest" by the authors. It turned out that significant differences were observed. Columns made from the nearly pure TEG-CD batch 1 only separated α-HCH and heptachlor *endo*-epoxide into enantiomers. Batch 2A was more suitable for polychlorinated pesticides. Besides an excellent enantiomer resolution of the two compounds mentioned before, U82 and *trans*-chlordane were also resolved. However, *o,p'*-DDT, *cis*-chlordane and toxaphene B9–1679 (Parlar50) were not, or only just slightly, separated. By far the best test results were obtained with batch 2B. With the exception of *cis*-chlordane, all compounds of the "chirotest" mixture were well resolved into enantiomers including *o,p'*-DDT (resolution $R =$ 0.89). On TEG-CD batch 1 *trans*-chlordane eluted before *cis*-chlordane. *trans*-Chlordane enantiomers are more retained with increasing under-ethylation and elute then after *cis*-chlordane.

We, as authors of the present monograph, tend to draw the following conclusion from the results thus far available in the literature: reproducibility of the procedures used for derivatisation of the cyclodextrins appears to be an important factor. Since in our laboratories and that of König the same methods are used, we have not encountered severe problems with different batches with regard to varying selectivities. The variability thus far encountered was acceptable. On the other hand, the results reported by Jaus and Oehme seem to indicate that the purity of the chiral selector is not necessarily a crucial parameter for *optimum* selectivity; however, very different performances of columns may be encountered when applying batches of the same material, but different derivatisation grades.

2.2
Enantioselective High Performance Liquid Chromatography (HPLC)

Resolutions can be divided rather arbitrarily into two major types, direct and indirect, as already outlined in the previous section. Here we will focus on the direct approaches and only briefly discuss the indirect methods.

2.2.1
Indirect Methods

Indirect methods are based on the reaction of a racemic mixture with a chiral reagent to form a pair of diastereomers. For example, a mixture of organic acid enantiomers reacts with an optically active organic base to give two theoretically separable diastereomeric salts (Fig. 2.2). Furthermore, the acid chloride of (*S*)-(–)-*N*-trifluoroacetylproline is a widely used derivatising agent and has been applied to the resolution of racemic amines and alcohols by HPLC [and also gas chromatography (GC)] [10]. Once the diastereomers have been resolved, the

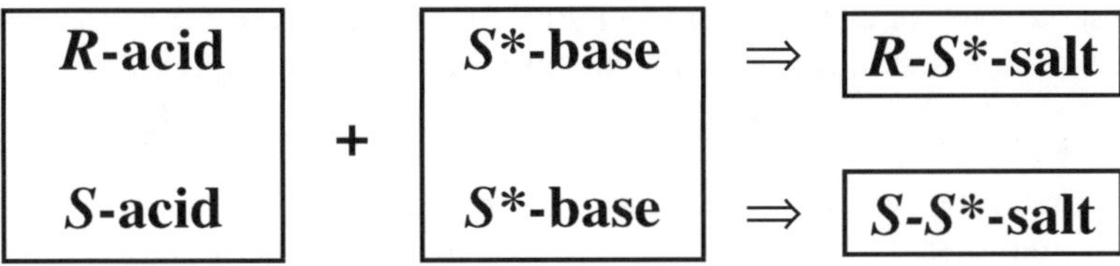

Fig. 2.2. Resolution of a racemic mixture of organic acid enantiomers into enantiomers by reversible formation of diastereomeric salts

enantiomers can be recovered by reversing the derivatisation procedure, e.g., by hydrolysis of the diastereomeric amides or esters. Although such indirect methods have been used extensively in the past in analysis, and are still being used on the process scale, they have a number of serious disadvantages. Therefore, Taylor and Maher conclude [10] *that an experienced analyst would generally prefer, if possible, to use a direct method of analysis.*

2.2.2
Direct Methods

Direct resolution of enantiomers is possible using chromatographic techniques, provided that chiral recognition is achieved by an optically active moiety, termed a *chiral selector*. This chiral selector must associate preferentially with one of the enantiomers in the racemic mixture; it may be part of the stationary phase or, alternatively, be present as a mobile phase additive, in which case an achiral stationary phase can be used (for Refs. see [10]). Many racemic mixtures have been separated on conventional achiral LC columns by adding an appropriate chiral compound to the mobile phase. Additives such as camphorsulphonic acid and quinine have been successful; the application of α-, β-, and γ-cyclodextrins has also been reported (for Refs. see [10]). Advantages of this technique are as follows:
- Less expensive conventional packings such as silica or octadecylsilica (ODS) can be used;
- there is a wide variety of possible additives; and
- the selectivities are sometimes different from those arising on the available chiral phases.

On the other hand, some disadvantages have to be noted:
- Many additives are costly or must be synthesised, and their mode of operation is complex;
- for preparative applications, the chiral additive must be removed from the enantiomeric solutes; and
- the detection mode may limit the choice of additives.

For these reasons, Taylor and Maher conclude [10] that the trend is to use chiral stationary phases whenever HPLC is considered for enantiomeric separations.

The separation of enantiomers by HPLC using chiral stationary phases (CSP) is based on the formation of transient diastereomeric complexes between the enantiomorphs of the solute and a chiral selector that is an integral part of the stationary phase. The difference in stability between these complexes leads to a difference in retention time: the enantiomer that forms the less stable complex will be eluted first. These diastereomeric adsorbates must, therefore, differ adequately in free energy for enantiomer separation to be observed.

2.2.3
The Evolution of Chiral Stationary Phases for Liquid Chromatography

The earliest review of chiral stationary phases for modern LC was published in 1984 [11], and was followed by several other recommendable overviews including [12–17]. In 1952, long before HPLC had emerged as the powerful instrumental separation method familiar to us now, an English scientist named Dalgliesh was studying the separation of amino acid enantiomers by paper chromatography [18]. Although not the first to observe such separations [19, 20], he correctly attributed them to enantioselective adsorption by the optically active cellulose molecules of paper, and proposed tentative requirements for such process:
- three points of attachment are required for stereochemical specifity in adsorption, and
- within the amino acid analyte to be resolved, the amino group, the carboxylic acid group, and a structural component of the side chain (such as the hydroxyphenyl group of tyrosine), should all participate in the molecular interactions, envisaged to be due to hydrogen bonding or steric repulsion.

Although Dalgliesh's studies were not unique and had been preceded by enantioselective adsorption on columns of materials ranging from powdered D- and L-quartz [21] to lactose [22], they are often quoted because of their formulation of the structural requirements for asymmetric recognition which are already very close to the well-known *Pirkle three point rule* [37]: "*Chiral recognition requires a minimum of three simultaneous interactions between the CSP and at least one of the enantiomers, with at least one of these interactions being stereochemically dependent*" (see below and Fig. 2.3). That means, at least one of the interactions will be absent by replacing one enantiomer with its antipode.

These early experiments were followed by a period of slow development, mainly in Europe, during which the types of materials used for column packings were extended to include starch [23–25], cellulose acetate [25], microcrystalline cellulose [26], Sephadex (cross-linked dextran) [27], surface-coated alumina [28], and surface-coated silica gel [29]. During this period, a number of synthetic polymers containing covalently bound asymmetric units were studied and found to yield partial or, in some cases, almost complete resolution of enantiomers. Examples of significance, reported mainly from German laboratories, include:
- a chiral ion-exchanger resin made by reacting quinine with cross-linked polyacryloyl chloride [30],

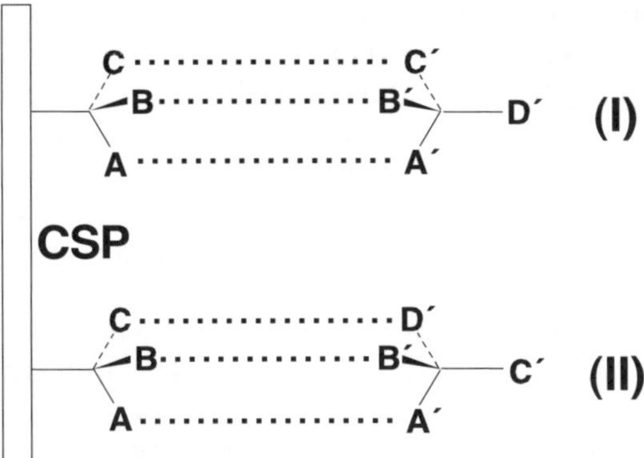

Fig. 2.3. Three-point interaction model of chiral recognition according to the Pirkle rule [37]; *CSP*: Chiral stationary phase ≡ 0 chiral selector; *I* and *II*: analyte enantiomers I and II

- the use of amino acid bonded polystyrene [31], and
- the application of polymers of asymmetric acryloyl amides [32, 33] as chiral adsorbents.

This early period has been well reviewed by Buss and Vermeulan [34] and by Davankov [35].

Although the number of publications about enantiomeric separations and the number of commercially available chiral stationary phases have expanded substantially, relatively few basic types of chiral stationary phases are available. Indeed, most of the enantiomeric separations that can be performed by LC can be accomplished using six or seven different columns [10, 17]. Several authors have tried to categorise HPLC-CSP into distinct classes, see for example [10, 17, 36]. In one of the systems developed by Wainer [36], the chiral recognition process is divided into two interdependent stages:
- the formation of the solute-CSP complex, represented by interactions A⋯A' and B⋯B' in Fig. 2.3, and
- the expression of the stereochemical differences between the enantiomorphs represented by the interactions C⋯C' and C⋯D', respectively, in Fig. 2.3.

The mechanisms involved in the first stage are used as a criterion for the division of HPLC-CSP into groups.

In detail, the process of chiral recognition can be resolved into the following steps [10, 37]: there exist three possible points of interaction between the chiral solute (sites A', B', and C') and the chiral selector bound to the chiral stationary phase (sites A, B, and C). Enantiomer *I* interacts at three sites A⋯A', B⋯B', and C⋯C', whereas its mirror image, enantiomer *II*, does not interact at site C⋯C'. If the C⋯C' interaction results in the stabilisation of the transient diastereomeric complex between enantiomer *I* and the *CSP*, enantiomer *I* will be retained on the column longer than enantiomer *II*. However, a destabilising C⋯C' interaction will reverse the elution order, i.e., enantiomer *I* will be eluted before enantiomer *II*. If the C$\cdots$C' interaction is minimal or non-existent, chiral recognition will not occur and the racemate will not be resolved. The three-point model suggests that at least three attractive interactions between the receptor or chiral selector and one of the enantiomers of the racemate need to be possible to obtain enantioselectivity. At least one interaction must depend on the stereochemistry at the chiral centre of both receptor and the enantiomer, otherwise unfavourable associations might be formed.

However, while the mechanism of operation of such phases remains unclear, Taylor and Maher suggested another simpler classification [10]:

– *Type 1*: The solute is part of a diastereomeric metal complex [chiral ligand exchange chromatography (LEC)]. This is a special case of ion exchange, and involves the reversible formation of a metal complex by coordination of substrates that can act as ligands to the metal ion; it has wider application than resolution of racemates. For further details the reader should refer to [10, 15, 17, 38, 39].

– *Type 2*: Essentially the only discernible interactions are attractive charge-transfer interactions. Pioneering studies were carried out by a team in Israel led by Gil-Av [40] who used packings designed exclusively for charge transfer complexation. Their objective was the resolution of helicene enantiomers, a particularly difficult objective in that neither acidic nor basic functions are present on the asymmetric molecule to assist with chiral recognition. Gil-Av's team achieved the enantiomeric resolution of this chiral compound by bonding a chiral charge-transfer acceptor, tetranitro-9-fluorenylideneaminooxypropanoic acid (TAPA) to aminopropylated silica gel. Such packings, as well as others which have meanwhile been developed, although they represent a pioneering achievement, have not proved general enough for most applications.

  The most recent version of the π-complex chiral stationary phases contain both π-acid and π-base groups. These chiral stationary phases tend to be more widely applicable than earlier phases of *Type 2* [63, 64]. Multimodal, multimechanistic chiral stationary phases that use both π-complexation and inclusion complexation seem to be very widely applicable [65], including applications to preparative LC. Further information on this *Type 2* phase as well as additional references can be found in [10, 17, 40–42].

– *Type 3*: The solute-CSP complexes are formed by multiple attractive interactions, including hydrogen bonding, π-π interactions, dipole stacking, etc., between the solute and the low molecular weight CSP. These are often termed *Pirkle-CSP*, after their principal inventor, or asymmetric strand CSP.

  During the 1980s, several groups reported generally useful packings in which relatively simple structures, containing just one or two centres of asymmetry, are chemically bonded to silica. This type of CSP is also termed variously "multiple interaction" [43] or "brush-type" [44], because their organic groups are directed away from the silica network in a similar way to the bristles of a brush. They are typically well-defined chemical structures, almost always containing at least one of the following functions near the chiral centre: (1) π-acidic or π-basic aromatic group, capable of donor-acceptor interactions (as in charge-transfer complexation); (2) polar hydrogen-bond donor/acceptor; (3) dipolar bond, suitable for dipole-dipole interactions as in dipole stacking; and (4) bulky nonpolar groups, providing potential for steric repulsion, van der Waals's interaction, and/or conformational control. Evidently, the solutes to be separated must contain similar but complementary groups if favourable discrimination is to be achieved. Despite the structural simplicity of such CSPs, a very large number of analytes have been resolved on them, and the world record α-value for chiral separations is held by one of this class [45]. Certainly an extremely wide variety of types of analytes have proved to be separable on such CSPs, including carbinols, hydantoins, lactams, succinimides, phthalides, sulfoxides, and sulfides [46]. If the analyte lacks all the necessary interaction sites, it may be possible to increase those present by analyte derivatisation, in this case using chiral reagents [47, 48]. Further appreciation of *Type 3* phases including additional references covering this field is given in [10].

– *Type 4*: Interactions between the analyte and CSP follow from inclusion into chiral cavities within a CSP of relatively low molecular mass, i.e., not a high molecular weight polymer. This effect was first demonstrated by Cram and co-workers [49] in Los Angeles using a chiral crown ether bound to silica gel. Although commercially available chiral crown ether stationary phases such as *Diacel's Crown-Pak* are still finding applications [50], attention has focused increasingly on the use of bound cyclodextrins [51, 52] for this type of CSP. The various moieties applied to functionalise cyclodextrins can alter their enantioselectivity greatly, thereby expanding their overall usefulness. For example, aromatic functionalised cyclodextrins can be used in the normal-phase mode as π-complex chiral stationary phases or in the reversed-phase mode where inclusion complexation dominates [66]. Completely different types of chiral molecules are resolved in each mode. Hydroxypropyl-functionalised cyclodextrins provide the best means of resolving *N-tert*-butoxycarbonylamino acids and various other chiral molecules [66–68].

A different experimental approach enhances the usefulness of native cyclodextrin stationary phases and produces unusual enantioselectivities. In this approach, the inclusion complexation is suppressed by using a non-hydrogen-bonding, polar organic solvent, such as acetonitrile, as the main component of the mobile phase [69–73]. Although this technique is sometimes called the "polar organic mode", it is related most closely to normal-phase separations. The acetonitrile tends to occupy the cyclodextrin cavity. It also accentuates hydrogen bonding between the hydroxyl groups on the chiral analyte [72–73]. The addition of a hydrogen-bonding solvent, such as methanol, can decrease the retention of highly retained compounds. Adding very small amounts of glacial acetic acid and triethylamine controls the protonation of the analyte and enhances enantioselectivity [72–73]. In this mode, the analyte presumably resides on top of the cyclodextrin selector – more or less a lid – in such a way that hydrogen bonding is maximised. Using this technique, chiral compounds can often be resolved containing two hydrogen-bonding groups (one of which should be α or β to the chiral centre) and a bulky group, such as an aromatic ring. This technique is also excellent for preparative separations. According to Armstrong [17], the polar organic mode offers some practical advantages over the traditional normal-phase mode, which uses n-hexane or n-heptane and n-propanol solvent systems. Firstly, most compounds that exist as hydrochloride salts can be dissolved and separated in the polar organic mobile phase, but not with non-polar organic solvents. Secondly, eluates from a reversed-phase column can be switched directly onto a chiral column when using achiral-chiral coupled column systems, as long as workers are using a polar organic-type mobile phase. This switch cannot be performed when operating a chiral column with traditional non-polar solvents, because they are incompatible with aqueous, buffer-containing solvents.

- *Type 5*: The primary mechanism for the formation of solute–CSP complexes is through attractive interactions, but inclusions also play an important role, because these CSPs are high molecular mass polymers. Some of these polymers are essentially of natural origin, such as those derived from cellulose and macrocyclic antibiotics, and others, which are synthetic in origin, are mostly based on polyacrylates or -methacrylates and polyacrylamides. Attempts to improve the chromatographic and enantioselective properties of cellulose have concentrated on derivatisation of its hydroxyl groups to decrease bulk polarity and to provide additional bulk for interaction between the CSP and the analyte. Hesse and Hagel [53] prepared microcrystalline cellulose triacetate (CTA) and observed that derivatisation of the hydroxyl groups did not destroy the helical structure of cellulose, which retained its potential for chiral recognition first noted in the early experiments of Dalgliesh [18] and others. Mannschreck and co-workers [54–56] introduced microcrystalline CTA for HPLC separations and, for example, resolved racemic compounds carrying aromatic groups. Certainly CTA can be used as a bulk packing, especially when cross-linked, but its pressure resistance is limited to around 80 bar. A report by Rizzi [57] claimed that its usefulness can be improved significantly by coupling a swollen microcrystalline CTA column with an achiral alkyl-silica column. The first successful commercial packings of this type consisted of silica-supported cellulose triacetate. Its versatility led Okamoto's group to develop a range of such phases by changing the derivatisation of cellulose [58]. These CSPs, immobilised on wide-pore silica gel, are now marketed under the tradename Chiralcel by the Japanese company Diacel Industries. They can be used with mobile phases such as pure ethanol and aqueous methanol and have been successfully applied to the separation of a very wide range of analytes. Although developed and commercialised entirely in Japan, they have been evaluated extensively in laboratories all over the world, and many publications are based on investigations carried out with these *Type 5* phases. More details can be found in [10].

Macrocyclic antibiotics are the newest class of chiral selectors applied to HPLC [74–78]. Two types of glycopeptide-based chiral stationary phases are commercially available: vancomycin and teicoplanin. Avoparcin will probably be the next available member of this class of chiral selectors. The glycopeptide macrocyclic chiral selectors appear to have broad applicability. They are multimodal chiral stationary phases because they can be used effectively in the reversed-phase, normal-phase and polar organic modes. The enantioselectivity is usually different in each mode [74–78]. These chiral stationary phases are useful for preparative separations as well.

Armstrong points out that the principle of complementary separations is a very useful concept when developing methods with glycopeptide-based chiral stationary phases [17]. Vancomycin and teicoplanin are similar, closely related chiral selectors. They can exhibit similar but different enantioselectivities. Consequently, if a partial separation is obtained on one chiral stationary phase, in many cases chromatographers can take the related column and obtain a baseline separation using identical or very similar conditions.

– *Type 6*: The CSP is a protein and the solute/CSP complexes are based on combinations of hydrophobic and polar interactions. This group may be regarded as a major subdivison of *Type 5*, as a globular protein is certainly a high molecular mass polymer offering opportunities for inclusion. *Type 6* HPLC phases belong to the most attractive types of chiral stationary phases for pharmaceutical applications. Although early examples of this class of CSP, based on either bovine serum albumin [59, 60] or acid glycoprotein [61], suffered from a lack of reproducibility and a tendency to deteriorate in use, the more recent commercial examples, such as the second-generation silica-bound α_1-acid glycoprotein (AGP) [62], appear to be more reliable and popular, although they are still fairly expensive. Further details can be inferred from [10].

2.2.4
Other Experimental Approaches for Enantioselective Liquid Chromatography

Hinze et al. [150] used chiral surfactants as chiral mobile phase additives, in order to enhance the resolution of enantiomers. Although surfactants have been previously employed in many different separation science applications, including chromatographic mobile phases, reports of chiral mobile phase additives for the liquid chromatographic separation of optical isomers are scarce. Hinze et al. attributed this to the fact that the synthesis of chiral analogues of the typical micellar-forming surfactants is not an easy task. They, therefore, suggested commercially available bile salt systems or mixed micellar aggregates composed of a conventional achiral surfactant together with a chiral surfactant. Most optical separations were affected employing C-18, 5 μm spherical packing material; the bile salt surfactants included sodium cholate, sodium deoxycholate and sodium taurocholate, chiral non-ionic surfactants used comprised digitonin, dodecyl-β-D-maltoside, octyl-α-D-glucopyranoside, and octyl-β-D-glucopyranoside. The usage of bile salts for resolution improvement may prove to be very beneficial. However, Hinze et al. pointed out that the presence of an alcohol such as 1-pentanol in the bile salt mobile phase is required for separation. This is most proba-

Table 2.1. Enantiomeric ratios [ER=(+)-/(−)-α-HCH] observed in the course of a liquid membrane experiment during a period of 14 days; Experiment 1: 1 mM *rac-α*-HCH/application of two filter papers; experiment 2: 1 mM *rac-α*-HCH/application of four filter papers; experiment 3: 0.5 mM *rac-α*-HCH/application of two glass fibre filters [152] (for a general definition of ER, see Sect. 3.1.2)

Experimental period (days)	Enantiomeric ratios		
	Experiment 1	Experiment 2	Experiment 3
1	–	1.33	–
2	–	1.08	–
3	–	1.08	1.08
4	1.22	–	1.27
5	1.27	–	1.00
6	1.44	1.56	1.00
7	–	1.27	–
8	–	1.33	–
9	1.17	1.33	–
12	1.17	1.22	–
14	1.17	–	–

bly due to the fact that the added alcohol improves the chromatographic efficiency of this mobile phase. Alternatively, the added alcohol may promote formation of liquid crystalline aggregates, which could favour chiral discrimination.

Furthermore, Hinze et al. [150] conducted liquid membrane separations using an experimental protocol similar to that reported for cyclodextrin membranes [151]. The supported membrane was formed by dipping the cellulose or other filter paper into an aqueous solution containing the chiral OIM-6 ionene. A "blank" membrane was also prepared in an analogous manner by dipping the paper in a solution that did not contain the ionene. The experiments were conducted by placing 10 mL of diethyl ether containing the racemate of the chiral compound in the left-hand side of the chamber and 10 mL of neat diethyl ether on the opposite side. Both sides were stirred at about 30 rpm with magnet stirrers. The amount of each isomer that permeated the supported liquid membrane was determined by HPLC analysis. The experiment was subsequently also carried out in a similar fashion using the "blank" membrane.

A slightly modified approach was used by Armstrong [12] and Möller [152] who employed cyclodextrins as mobile carriers in an aqueous phase. Applying this technique, Möller was the first to investigate the enantiomeric separation of the environmental pollutant α-HCH. In this case, the supported membrane was formed by dipping a filter paper (Schleicher & Schüll, Germany) into an aqueous solution containing β-cyclodextrin (0.7 M). The mobile phase consisted of an aqueous 0.7 M β-cyclodextrin solution, 125 mM urea, and 37.5 mM NaOH. The latter two compounds were added in order to improve the solubility of β-cyclodextrin in water. The results obtained with 1 mM and 0.5 mM *rac-α*-HCH solutions during an experimental period of 14 days are summarised in Table 2.1. The most effective enantiomeric separations were achieved with the experimental approach used in *Experiment 2*, followed by *Experiment 1*, while the application of glass fibre filters turned out to be less successful. Furthermore, it is worth noting that a preferential permeation of (+)-*α*-HCH through the liquid membrane was observed by Möller, where the maximum enantiomeric ratio was encountered after about 6 days. As Möller aimed at a complete enantiomeric separation, which was not accessible by this technique, no further optimisation was carried out. Therefore, the actual potential of the liquid membrane technique is not yet clear.

2.2.5
Liquid Chromatography as a Measurement Tool for Chiral Interactions

In a recent review article, Ringo and Evans [79] summarised the potential of liquid chromatography beyond usual applications such as separating and analysing components of complex mixtures: HPLC can be utilised as a tool for fundamentally measuring interaction chemistry. This paradigm shift results from the basic insight that a significant amount of quantitative and qualitative information is contained in a separation, i.e., the same physicochemical interactions that produce a separation in the spatial domain can be measured using that spatial resolution as a parameter. HPLC can be tailored to interaction measurements of practical and fundamental interest. During the almost five decades since the advent of liquid chromatography, a broad scope of physicochemical measurements has been realised, including the determination of binding constants, partition coefficients, and diffusional parameters as well as interaction and reaction kinetics [80–82].

Although enantiomers have identical chemical and physical properties in an isotropic environment, they often exhibit significant differences in interactions with other chiral species. These enantioselective interactions form the cornerstone for many processes of biological and technological importance and have profound implications for pharmacology, molecular biology, and bioengineering. In many cases, however, the absolute energetics of these interactions are quite small, and the differences of importance are even smaller. But, as separation methods are based on weak interactions, they are uniquely well suited to the difficult measurement of such interactions. Basically, a chiral selector is either immobilised on a surface support as a stationary phase or present as an additive to the mobile phase. In all cases, chromatographic parameters are determined for these chiral separations and then related to, for example, thermodynamic parameters governing these enantioselective interactions.

The correlation of separation parameters to the energetics of solute/phase interactions is possible because LC retention processes are well described by equilibrium thermodynamics [83]. Each solute zone proceeds through the column at a rate controlled by competing interactions of the solute with the stationary phase and the solute with the mobile phase. This retention process results in an

Table 2.2. Thermodynamic interaction parameters determined by HPLC [79]

Thermodynamic parameter	HPLC measurement
Equilibrium constant for complexation, K_{comp}	Solute capacity factor, k'; Phase ratio, Φ
Chiral selectivity, α	Solute capacity factor, k'
Complexation stoichiometry	Solute capacity factor, k'; Solute peak area
Change in enthalpy upon complexation, ΔH_{comp}	Temperature dependence of solute capacity factor, $k'(T)$
Change in entropy upon complexation, ΔS_{comp}	Temperature dependence of solute capacity factor, $k'(T)$
Change in volume upon complexation, ΔV_{comp}	Pressure dependence of solute capacity factor, $k'(P)$

increase in the solute migration time (t_R) relative to the movement of a nonretained species (t_0) and is often described by the solute capacity factor (k'), which is equal to $(t_R-t_0)/t_0$. Although mobile-phase flow creates an inherently non-equilibrium condition, LC retention has been shown to be well modelled as an equilibrium process. Ringo and Evans [79] attribute this to the fact that solute retention is assessed from movement at the centre of the zone profile, where non-equilibrium effects from mobile-phase flow are minimised. As a result, a capacity factor measurement in a flowing HPLC system may be directly correlated to the equilibrium thermodynamics of solute interactions with the stationary phase and mobile phase. As shown in Table 2.2, a considerable range of thermodynamic parameters is thus accessible using LC measurements [83]. For an extensive discussion including a derivation of the respective equations for calculation of these parameters the reader should refer to [79] and the literature cited therein.

Furthermore, Ringo and Evans [79] emphasise that, in addition to quantitative measurements of enantioselective binding, LC can also be used for qualitative investigations of solute binding interactions with the selector. Many large chiral selectors may exhibit several distinct binding sites, with proteins as the most prevalent example in the literature. The interactions of solutes with different binding sites can be independent, competitive, or allosteric, whereby the binding of a solute at one site perturbs the binding of a solute at a different site. The number and cooperativity of active sites involved in a complexation interaction play an important role in the measurement and interpretation of all the thermodynamic parameters shown in Table 2.2. This information also has considerable practical impact in the prediction of competitive binding between enantiomers as well as competitive and allosteric interactions with other solutes.

Several recent investigations have focused on enantiomeric binding sites on chiral selectors [84–86]. In these studies, an immobilised-selector stationary phase was initially equilibrated with a mobile phase containing a single enantiomer. The retention of the other enantiomeric solute injected onto the column under these conditions was then compared with the retention of the solute when no enantiomer had been dissolved in the mobile phase. If the enantiomers bind at the same location(s), the retention of the injected enantiomer will be decreased by the presence of the equilibrated enantiomer within the stationary phase. Moreover, if the column is equilibrated with a series of marker solutes that are known to bind at specific sites on the selector, the precise location of the binding site for each enantiomer can be systematically determined. Although in many cases the structure of these binding sites remains ambiguous, the information gained from these studies is valuable when used in conjunction with other structural or functional data.

2.2.6
Recent Reviews on Enantioselective HPLC

Reviews that appeared in the year 1999 on enantioselective HPLC include [87] (advantages and limits of commonly used chiral stationary phases; 8 refs), [88] (library of 50 dipeptide 3,5-dinitrobenzoyl CSPs; results of screening of the library for the separation of a test racemate; 18 Refs.), [89] (enantioseparations of cardiovascular drugs; 200 Refs.), [90] (high performance and high selectivity packing materials; 53 Refs.), [91] (application to clinical samples; 243 Refs.), [92] (biologically active compounds of a cyanobacterium; 14 Refs.), and [93] (scaling-up procedures on chiral separations by HPLC; quick preparation of enantiomers in the developing stage of new drugs; 14 Refs.).

2.3
Enantioselective High Resolution Gas Chromatography (HRGC)

Present possibilities and limitations associated with enantioselective high resolution gas chromatography (HRGC) are summarised in several monographs and review articles. A general survey is given in the monographs by König [1, 133] and in review articles by Schurig and Nowotny [116] and Schurig [132], while the aspect "enantioselective gas chromatography of chiral environmental pollutants" is addressed in review articles by Hühnerfuss and Kallenborn [94], Hühnerfuss et al. [95], Vetter and Schurig [96], and Hühnerfuss [97–99]. An extensive review of chiral stationary phases known until 1982 has also been given by Liu and Ku [128].

2.3.1
The Evolution of Chiral Stationary Phases for Capillary Gas Chromatography

Gil-Av et al. were the first to report a successful separation of *N*-trifluoroacetylated (*N*-TFA) amino acid esters by capillary gas chromatography applying acylated amino acid and dipeptide esters as chiral stationary phases in glass capillary columns [100, 101]. Under such experimental conditions, D-enantiomers eluted prior to L-enantiomers from stationary phases containing or consisting of

L-amino acids. The separation performance of these early phases was largely attributed to hydrogen-bonding interaction between the enantiomers of the sample and the chiral stationary phase, where the diastereomeric association complexes between molecules of equal configuration were assumed to be more stable, leading to longer retention times, than in the case of unequal configuration. Particularly large separation factors for amino acids were observed in the case of the dipeptide derivative *N*-TFA-L-valyl-L-valine cyclohexyl ester. However, a serious disadvantage of this phase was the fact that the maximum temperature of operation was limited to 383 K (110 °C) only. The range of application could be extended by introducing other amino acids, e.g., L-leucine, L-aspartic acid, or L-phenylalanine, into the dipeptide phases [102, 103]. Another type of chiral stationary phase suggested by Gil-Av and Feibush [104, 105] consisted of carbonyl-bisamino acid esters, so-called "ureido phases", which offered the possibility of separating the enantiomers of acylated secondary amines in addition to amino acids. Particularly high enantioselectivity was achieved by *n*-dodecanoyl-L-valine-*tert*-butylamide phases introduced by Feibush in 1971 [106]. This diamide possessed only one chiral centre; however, several centres were suitable for the formation of hydrogen bonds. Somewhat unexpected was the wide range of application of *n*-dodecanoyl-(*S*)-α-(1-naphthyl)ethylamide with only one amide function [107]. On this phase, not only amino acids and amines, but also α-chiral carboxylic acid amides could be separated.

According to König [1], the general principle of all these separations appears to be the presence of nitrogen atoms in the substrate molecules. The wish to extend the range of application of enantioselective gas chromatography to other substrates led to another concept, i.e., the idea to adapt the structure of the stationary phase to the structure of the substrates to be separated. Along this line, the separation of acylated α-hydroxy acid esters was accomplished on (*S*)-α-hydroxyalkanoic acid-(*S*)-α-phenylethylamides [108–110]. The insufficient thermostability of these phases, however, again limited broader application of this type of phase. Furthermore, enantiomeric separation of amino acids, some α-hydroxy acids, alcohols, nitriles, and lactones was also achieved using the low molecular weight stationary phases developed by Ôi et al. [111–113].

A considerable advance in the applicability and practicability of enantioselective gas chromatography represented the fixation of the chiral selectors to a polymer, a very innovative new concept suggested by Frank et al. at the end of the 1970s [114, 115]. By copolymerisation of dimethylsiloxane with (2-carboxypropyl)methylsiloxane and coupling of the carboxy group with the amino group of L-valine-*tert*-butylamide, a polysiloxane with chiral side chains was obtained. This chiral polymer, which is now marketed as Chirasil-val®, exhibited both increased temperature stability above 473 K (200 °C), excellent enantioselectivity, and a wide scope of application (for reviews, see [117–119]). In line with this concept, however, a different approach was pursued by Verzele and co-workers [120, 121]: after hydrolysis of the cyanoalkyl side chains of the polysiloxanes OV-225 or Silar-10C, they obtained chiral polymers by coupling L-valine-*tert*-butylamide to the carboxyl groups. However, according to König [1], temperature stability and selectivity towards amino acid enantiomers were not quite as good as in the case of Chirasil-val®.

Similarly, König and Benecke [122] functionalised cyanoalkyl-polysiloxanes by reducing the nitrile group to aminomethyl groups and coupling N-acylated amino acids. These phases were well suited to separate amino acids, amino alcohols, and amines. Far more versatile in their applications turned out to be chiral polymers prepared by modifying the polysiloxane XE-60 [110, 123–125]. Fused-silica capillary columns with Chirasil-val® and XE-60-L-valine-(S)-α-phenylethylamide have become commercially available [126].

An interesting new route to chiral polymers was introduced by Schomburg et al. [127]: Carbowax 20M and acryloyl-L-valine-(S)-α-phenylethylamide were cross-linked by treatment with dicumoyl peroxide and at the same time immobilised inside the capillary column. The resulting coating is suitable for the separation of amino acid enantiomers, but – according to König [1] – the separation factors are lower than those observed for XE-60-L-valine-(S)-α-phenylethylamide. Further details, including its preparation, on the latter phase can be found in [1]. In all these cases, the chiral selectors are covalently connected to a methyl polysiloxane matrix, thus increasing thermal stability; however, the stereoselective interaction of the enantiomers with the chiral selector is based on hydrogen-bonding forces. Therefore, the ability to separate enantiomers using these phases is limited, with few exceptions, to substrates with hydrogen-bonding donor or acceptor functions.

Diastereomeric association may also take place between chiral molecules and chiral transition metal complexes. An experimental approach taking advantage of this principle for enantioselective separation was first described by Schurig [129, 130]. Since hydrogen bonding is not essential for chiral recognition in these complexes, a number of compounds without functional groups of this kind could be separated, such as cyclic ethers, epoxides, and spiroacetals. Thus, complexation gas chromatography can be considered complementary to enantiomeric separation with diamide phases.

In 1983, Koscielski et al. were the first to report gas chromatographic enantiomeric separations using cyclic glucose oligomers, so-called cyclodextrins [131]. Because of the overwhelming scope of application of chiral stationary phases with modified cyclodextrins, in particular, with regard to the central subject of the present volume, i.e., chiral environmental pollutants, this experimental approach deserves a more extensive description.

Cyclodextrins are cyclic α-(1→4)-connected glucose oligomers with 6, 7, or 8 glucose units, corresponding to α-, β-, and γ-cyclodextrin, respectively (Fig. 2.4). They can be prepared by enzymatic degradation of starch with cyclodextrin-glucanosyltransferase (CGTase) from *Bacillus macerans*, *B. megaterium*, and other bacterial strains [143], but they are also commercially available. A total chemical synthesis of α-cyclodextrin and of a "mannose isomer" has also been reported in the literature. For further details the reader is referred to [133, 143].

As shown in Fig. 2.4, the cyclodextrins exhibit a torus-shaped geometry with specific dimensions of their cavity, with the 6-hydroxy groups positioned at the

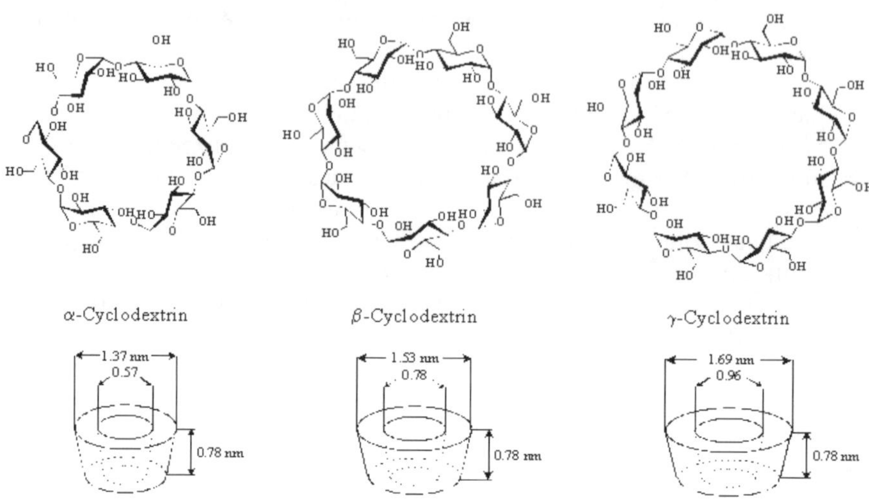

Fig. 2.4. Chemical structure and molecular dimensions of α-, β-, and γ-cyclodextrin [133]

narrow entrance of the cavity and with the 2- and 3-hydroxy groups at the wider opening. Their overall conformation is mainly determined by the α-(1→4)-connected glucose units in their $^4C_1$-conformation and stabilised by intramolecular hydrogen-bonding forces between the 2- and 3-hydroxy groups. In contrast to the hydrophilic character of the outer surface of the cyclodextrins, the interior of the cavity is hydrophobic and favours the selective inclusion and trapping of nonpolar guest compounds. However, gas chromatographic efficiency in terms of the number of theoretical plates is poor with unmodified cyclodextrins irrespective of whether they are used as solids (gas-solid chromatography) or in formamide solution. Improved peak shapes were achieved when per-O-methylated cyclodextrins were used in capillary columns, as described by Szejtli and co-workers [134, 135], although the disadvantage of these phases was their unfavourably high melting point. As a consequence, different experimental approaches were used to lower the melting point. For example, Venema and Tolsma [136] proved that per-O-methylated β-cyclodextrin behaves as a supercooled liquid after conditioning at 473 K (200 °C), and afterwards the columns could be used as low as 349 K (76 °C). Schurig and co-workers diluted the cyclodextrin derivative in a polysiloxane, e.g., OV-1701 [137]. König, Wenz and co-workers [138, 139] successfully applied per-O-pentylated and selectively 3-O-acylated-2,6-di-O-pentylated α-, β-, and γ-cyclodextrins. Most of these derivatives are liquid at room temperature, highly stable, soluble in nonpolar solvents and with high enantioselectivity towards many chiral compounds. Subsequently, a considerable number of hydrophobic cyclodextrin derivatives was prepared and evaluated. For an extensive review of these more recent achievements the reader is referred to [116, 132, 133]. The development of these latter chiral stationary phases represents a landmark in the enantioselective analysis of chiral environmental pollutants, because they also allow the enantiomeric separation of hydrophobic compounds such as α-hexachlorocyclohexane, chlordane, heptachlor, polychlorinated biphenyls, and toxaphenes, etc. Most of the investigations reported in Chapter 3 have been and are being carried out with modified cyclodextrins of these types.

2.3.2
Enantioselective Multidimensional Capillary Gas Chromatography (MDcGC)

A multidimensional gas chromatography (MDGC) technique was suggested by Schomburg et al. [140], as well as by Duinker and co-workers [141, 142], where the latter group aimed at an effective separation of PCB congeners which at that time could not be separated on a single SE-54 or comparable columns by conventional methods. The MDGC technique involves the application of two columns of different polarities in series, each in a separate temperature-controlled oven. The eluate of the first column is carried either through the monitor ECD (ECD ≡ Electron capture detection), producing the usual ECD chromatogram, or through the second column and the main ECD. The remarkable advantage of this technique, in both a qualitative and quantitative sense, is a consequence of the fact that a valveless pneumatic system involving a live T-piece allows a preselected small fraction to be cut from the eluate of the first column and to be transferred quantitatively and reproducibly to the second column. This technique offers complete separation of many compounds coeluting on a single column, increased selectivity, and it supplies very accurate data. However, special two-oven gas chromatographs and skilled personnel are necessary to perform these time-consuming analyses.

Examples of successful applications of this technique with regard to enantiomeric separation of chiral environmental pollutants can be found in several publications by Schurig and co-workers [144–147, 149], de Geus et al. [192], Benicka et al. [191] and by Parlar and co-workers [148], the results of which are discussed in Chapter 3.

2.3.3
Other Experimental Approaches for Enantioselective Capillary Gas Chromatography

The gas chromatographic isomer selective separation of complex mixtures of environmental pollutants such as chlordanes or toxaphenes was and still is a challenge. Enantioselective separations at least double or even multiply this problem. One possible approach that may significantly improve separation in such difficult situations may be multidimensional gas chromatography (see Sect. 2.3.2), another possibility may be the application of tandem gas chromatography, i.e., direct coupling of two capillary columns with different stationary phases in series. However, the chance of finding an optimum tandem system is rather small if the column properties (e.g., phase, length, film thickness) are selected by a "trial-and-error" approach. Jones and Purnell developed a general theory which allows optimisation of the separation columns coupled in series, in particular, the calculation of the length of combination necessary for the baseline separation of a compound mixture. The theory was verified experimentally with complex mixtures, which could not be separated by single columns of any length (for refs see [189, 190]).

The first to suggest the tandem combination of two columns with an achiral and a chiral phase in order to enhance the separation of chiral environmental pollutants were Oehme et al. [153]. They reported the simultaneous enantioselective separation of chlordanes, a nonachlor compound, and *o,p'*-DDT in environmental samples using tandem capillary columns. Basically, the order *isomer-enantiomer separation* can be chosen or *vice versa*, but Oehme et al. suggested the first arrangement because of the retention time shift phenomenon which led to insufficient enantiomeric separation [153]. The drawback of the sequence isomer-enantiomer separation is the relatively high temperature at which the compounds are transferred from the isomer-selective column into the enantiomer-selective one. This may cause some deterioration of enantiomer resolution, which can be partly compensated by reduction of the upper column temperature applied. Furthermore, the theory of Jones and Purnell requires isothermal separation, which is not feasible for the separation of complex mixtures of environmental pollutants such as toxaphenes which show relatively small differences in their retention properties [190].

Therefore, Baycan-Keller and Oehme [189] attempted to overcome this limitation of the theory by approximating isothermal conditions by using a very fast heating ramp to reach the selected isothermal temperature as quickly as possible. This change had no influence on the predictability of the elution order by the theory. In addition, Baycan-Keller and Oehme presented calculation procedures that allow the prediction of the best possible isomer and enantiomer separation at the shortest possible tandem system. The latter argument was assumed to be important, because enantiomer separation is usually best at low temperatures. The procedure suggested by Baycan-Keller and Oehme only requires the measurements of the retention times and of the column hold up time of the single columns. The authors stress that for the determination of the optimum length of the single columns for the tandem system a substantial amount of calculations is necessary. However, they are mathematically trivial and simple computer programs can be used.

Additional examples of successful applications of tandem GC with regard to enantiomer separation of chiral environmental pollutants can be found in publications by Garrison et al. [162] who investigated the enantioselectivity, or lack of, occurring during the environmental degradation and biological exposure of *o,p'*-DDT and -DDD investigated by tandem GC and M-CEKC (see also Sect. 2.4.2). With regard to tandem GC, the combination of permethyl-trifluoro-acetoxypropyl-*γ*-cyclodextrin (Chiraldex G-PT) as chiral stationary phase and an achiral XTI-5 column (5% diphenyl, 95% dimethyl polysiloxane phase), in that order, provided excellent separation of the six main DDT isomers/congeners (*o,p'*-DDT, -DDD, and -DDE and *p,p'*-DDT, -DDD, and -DDE) as well as baseline separation of the enantiomers of *o,p'*-DDT and, for the needs of their experiments, of *o,p'*-DDD. Reversal of order of the columns provided little additional efficiency over that using the γ-cyclodextrin column alone. Although the enantiomeric separation was approximately the same with the chiral column alone as with the tandem columns, without the XTI-5 column one of the enantiomers of each of the two chiral compounds overlapped to the extent that it was impossible to determine whether the loss of *o,p'*-DDT or the formation of *o,p'*-DDD was enantioselective.

2.3.4
Possible Sources of Error of Enantioselective cGC

Potential sources of error of enantioselective capillary gas chromatography were extensively discussed by Schurig [193] and by Vetter and Schurig [96]. The most important points that should be considered are as follows:
- decomposition of the analyte at high temperatures during chromatography; this implies that the enantiomer which spends a longer time in the column will be lost preferentially causing an error in the enantiomeric ratios;
- coelution of impurities spuriously increasing peak areas; basically, this effect has also to be observed in the course of cGC with achiral stationary phases. It should be noted, however, that the application of chiral stationary phases doubles the number of peaks of chiral components and thus the chance of coelution effects;
- racemisation of labile enantiomers causing peak distortions due to inversion of configuration during enantiomer separation;
- peak distortions caused by inadequate instrumentation; and
- nonlinear detector response.

For a deeper discussion on these potential errors the reader is referred to the review paper by Vetter and Schurig [96]. Additional errors can be caused by unsatisfactory calibration curves, overloading phenomena and possible lack of standards of highest optical purity. This latter aspect will be pursued in Section 2.6.

Furthermore, Meyer pointed out that the chromatographic quantitation of very low amounts of an enantiomer in the presence of its antipode can be an extraordinary challenge [194]. If resolution of the peaks is not complete even at extreme mass ratios, an integrator will yield inaccurate results due to geometric effects. A given resolution can be adequate for peaks of similar size, but result in severe overlap, if one of the signals is markedly smaller. If tailing occurs, the problem appears to be especially severe for last-eluted small peaks.

2.3.5
Recent Reviews on Enantioselective HRGC

Reviews that appeared in the year 1999 on enantioselective HRGC include [87] (advantages and limitations of commonly used chiral stationary phases; 8 refs), and [92] (biologically active compounds of a cyanobacterium; 14 refs).

2.4
Capillary Electrophoresis (CE)

2.4.1
General Considerations

High-performance liquid chromatography and high-resolution capillary gas chromatography are still the dominating approaches used for the enantioselective separation of chiral environmental pollutants. However, enantiomer separation by capillary electrophoresis (CE) has rapidly attracted attention as a promising field of pharmaceutical sciences following the innovative work of Jorgenson and Lukacs [154] and Terabe et al. [155] in the early 1980s. In 1987, CE instruments became commercially available, largely stimulated by pharmaceutical, clinical and biomedical demands. For a brief introduction to this technique the reader may wish to consult monographs, such as [174–176], or review articles, for example, [156, 177, 178].

Separation can be achieved with one capillary tube very quickly with high resolution. Performing electrophoresis in small-diameter capillaries allows the use of very high electric fields, because the small capillaries efficiently dissipate the heat that is produced. Increasing the electric fields in turn produces very efficient separations and reduces separation times. Capillaries are typically of 50 μm inner diameter and 0.5 to 1.0 m in length. The applied potential is 20 to 30 kV. Helmholtz described the phenomenon that an electrical field may induce an electroosmotic flow within a capillary around 1850 [172]. Electroosmotic flow arises from two effects. Firstly, an induced electrical double layer occurs at the solid-liquid interface of the fused-silica capillary tube and the electrolyte. This double layer is induced by the presence of ionisable silanol groups at this interface. In the applied electric field, the positive counterions in the mobile phase are mobile and in slight excess, which causes a net movement of solvated cations towards the negative electrode. Secondly, a similar double layer occurs on the silica-based column packing. The positive counterions also cause this double layer to contribute to the net movement of the cations in an aqueous buffer mobile phase. Due to the electroosmotic flow, all sample components migrate to the negative electrode. A small volume of sample (about 10 nL) is injected at the positive end of the capillary and the separated components are detected near the negative end of the capillary. CE detection is similar to detectors used in HPLC, and includes absorbance, fluorescence, electrochemical, and mass spectrometry. The electrophoretic mobility of a substance is a function of the charge q and mass of the particles. Furthermore, the mobility depends on the dielectric constant ε and viscosity η of the solutions as well as of the strength of the electrical field E and the zeta-potential ζ [156]. Basically, the capillary can also be filled with a gel, which eliminates the electroosmotic flow. Separation is then accomplished as in conventional gel electrophoresis, but the capillary allows higher resolution, greater sensitivity, and on-line detection.

The versatility of CE is remarkable, because various experimental modes can be used which are either based on electrophoretic or chromatographic principles [156]. The most attractive modes include *capillary zone electrophoresis* (CZE), *capillary electrokinetic chromatography* (CEKC), *capillary gel electrophoresis* (CGE), *capillary isotachophoresis* (CITP), *capillary isoelectric focusing* (CIEF), and *capillary electrochromatography* (CEC) [157]. In both CZE and CITP, the separation is based on the differential mobilities of the analytes, while CEKC and CEC are based on chromatographic separation principles. CGE is closely related to molecular sieve effects and CIEF separations occur due to differential pK values. Accordingly, the different modes are preferentially used for specific applications: CZE for the analysis of small charged molecules, CEKC also for neutral analytes, CGE for nucleic acids, CIEF for the determination of the isoelectric point of proteins and for the separation of immunoglobulins or haemoglobins [157].

2.4.2
Enantioselective CE

As two enantiomers do not possess different charges, their separation cannot be achieved on the basis of "pure electrophoretic principles". In general, a chiral environment is required for enantioselective separations, and, therefore, the most common approach in CE is the addition of a chiral selector to the buffer solution. Due to the different stereoselective interactions between the enantiomers of the racemic analyte and the chiral selector a differential mobility is encountered. In the case of CE, enantioselective migration is accordingly dependent on electrophoretic principles, while the separation itself is largely attributed to a chromatographic mechanism [157]. The chiral selectors most often used in CE include cyclodextrins, natural micelle formers and macrocyclic antibiotics. It is important to note that the chiral selector, the analyte and their diastereomeric complexes must exhibit different mobility.

Of the various CE modes summarised above, CZE and CEKC, in which only a chiral selector is added to the usual running buffer solution, are most widely used for enantiomer separations (see [157] and refs cited therein). One of the most attractive advantages of CZE and CEKC is the easy change of separation media in the method development, i.e., the separation solution can easily be altered to find the optimum separation media and also an expensive chiral selector can be used because of the small amounts of media required. Electrically neutral chiral selectors are employed for the enantiomer separation of ionic solutes in CZE. Cyclodextrins have been found to be most effective in CZE for a wide range of ionic drugs. On the other hand, charged pseudo-stationary phases are added to the running buffer solution in CEKC. Therefore, CEKC can be applied to both ionic and non-ionic solutes.

Cyclodextrins are also as effective in micellar CEKC as in CZE. In his review article, Nishi focused on the enantiomer separation of drugs by different CEKC modes, mainly micellar CEKC (M-CEKC) and affinity CEKC (A-CEKC) [157]. In the latter mode, biological components such as proteins or mucopolysaccharides are employed as novel chiral selectors. CEKC, in which charged cyclodextrins were applied as chiral pseudo-stationary phases, have been successfully used for enantiomer separations of both ionic and non-ionic solutes. Diastereomeric analytes have been separated by CEKC as in reversed-phase HPLC. The same chiral derivatisation reagents as used for HPLC analysis and typically M-CEKC with SDS have been employed.

Recently, Schurig and co-workers [158] suggested an interesting variation of CEKC: dual enantiomer discrimination involving β-cyclodextrin derivatives in the mobile and the stationary phase. The application of anionic and cationic cyclodextrins, employed as chiral mobile phase additives, showed an opposite enantioselectivity compared with Chirasil-Dex. Thus, the chiral separation factor $\alpha > 1.5$ could be decreased or increased by the addition of these additives.

Examples of successful applications of CE with regard to enantiomer separation of chiral environmental pollutants can be found in several publications by Benito et al. [161] (separation of chiral PCBs by cyclodextrin-modified M-CEKC), Garrison et al. [162] (environmental degradation and biological exposure of o,p'-DDT and -DDD investigated by tandem GC, see Sect. 2.3.3, and M-CEKC), and Grainger et al. [163] (enantioselective separation of chiral PCBs and chiral phenoxy herbicides by cyclodextrin-modified C-MEKC with UV and MS detection).

2.4.3
Other Experimental Approaches for Enantioselective CE

The combination of capillary electrophoresis (CE) and HPLC packings is assumed to be one of the more exciting developments in the separation sciences [159]. As the principles of this relatively new method are not yet well known, a brief description is due: The basis of CE separations is the differential movement of neutral compounds through a packed capillary with an inner diameter of less than 100 µm using electroosmotic flow as the driving force. The advantage of using electroosmotic flow is increased column efficiency because of the plug-flow profile and the ability to use smaller particles than those used in HPLC. Hydraulic flow in HPLC is limited by the column back pressure that results from the passage of the mobile phase through a packed bed. In CE, electroosmotic flow is limited only by the voltage applied. The particle diameter of the packing has no major effect on electroosmotic flow. The minimum particle diameter is determined more by practical reasons – the smallest particles that can be obtained and the means to contain them in the fused-silica capillary. Although the technique is still in the early developmental stage, practitioners using CE have achieved isocratic separations displaying hundreds of thousands of theoretical plates with separation times comparable to those of HPLC.

An interesting parameter in CE is the nature of the packing material itself. Of course, the packing has a large influence on the separation [159], because the normal chromatographic interactions occur between the solute, the mobile phase and the stationary phase. In addition to the chromatographic interactions, the packing material has a profound effect on the electroosmotic flow itself. These differences in behaviour will presumably lead to studies of optimised packings for CE and perhaps to fused-silica deactivation procedures [159].

Hyphenation of CE to nano-scale NMR is another topic with considerable perspective. By coupling of NMR to CZE, a nanoliter-scale detection limit can be achieved. Compared with hyphenation of CZE in open tubular columns to electrospray mass spectrometry (ES-MS), the main advantage of this technique is that no interface is necessary for coupling, because of the higher flow rate of the mobile phase [160].

2.4.4
Recent Reviews on Enantioselective CE

Recent reviews on enantioselective CE include [156] (comprehensive survey; 51 refs; in German), [164] (electrochromatography utilising modified cyclodextrins as stationary phases; 69 refs), [165] (CE-MS technology; 62 refs), [166] (trends in capillary electromigration techniques; 62 refs), [167] (CE using macrocyclic antibiotics as chiral selector; 20 refs), [168] (temperature effects, temperature gradients; 51 refs), [169] (method development and validation for pharmaceutical and biological applications; 840 refs), [170] (enantiospecific bioanalysis; modern drug development; 63 refs), [171] (molecular imprint-based stationary phases for CE; 30 refs), and [173] (current applications in the analysis of pharmaceuticals; 47 refs).

2.5
Supercritical Fluid Chromatography (SFC)

2.5.1
General Considerations

Supercritical fluids can be used as mobile phases to separate analytes with chromatographic columns. As in supercritical fluid extraction (SFE), supercritical fluids can exhibit solvating potentials similar to organic solvents, but with higher diffusivities, lower viscosity, and lower surface tension. The lower viscosity allows higher flow rates compared with liquid chromatography, and the solvating power can be adjusted by changing the pressure. A major advantage of supercritical fluid chromatography (SFC) is that it derives benefit from the liquid-like solubility, with the capability to use a non-selective gas-phase detector such as a flame ionisation detector (FID). Analytes that cannot be vapourised for analysis by gas chromatography, yet have no functional groups for sensitive detection with the usual liquid chromatography detectors, can be separated and detected using SFC. The column is usually a capillary GC column, but packed LC columns can also be applied. The most common detector is the FID, but other GC or LC detectors can also be employed. Compared with capillary GC, the strongly enhanced solvation strength of, e.g., supercritical carbon dioxide, allows the oven temperature to be reduced significantly [179]. This, in turn, offers a promising alternative to increase the enantioselectivity of a chiral stationary phase, since the separation factor α increases with decreasing separation temperatures [180] in the usually enthalpy-controlled domain of enantiomer separations [181].

2.5.2
Enantioselective SFC

The increased solvation power of supercritical mobile phases necessitates the development of non-extractable inert stationary phases, e.g., chemically modified chiral polysiloxanes, which can be cross-linked and/or immobilised onto the surface of fused-silica capillaries, resulting in stable films with low bleeding rates and high wash-out resistance [182, 183]. For example, Schurig et al. [182] investigated chiral polysiloxanes containing chemically bonded permethylated β-cyclodextrin (Chirasil-Dex) or metal(II)-bis[3-heptafluorobutanoyl-camphorate] (Chirasil-METAL). The authors reported that the capacity factors k' were lower in SFC than in GC. Whereas for Chirasil-Dex the chiral separation factors α decreased with increasing mobile phase density, for Chirasil-NICKEL they slightly increased. Schurig et al. concluded that for Chirasil-Dex, low analysis temperatures combined with low densities yielded the best chiral resolution. For Chirasil-NICKEL, low temperatures were favourable, even if high densities were used. Examples of enantiomer separation using these chiral stationary phases include pharmaceuticals such as ibuprofen, ketoprofen, cicloprofen, flurbiprofen, the rodenticide warfarin, and many other compounds of various classes [182, 183, 200].

2.5.3
Recent Reviews on Enantioselective SFC

Recent reviews on enantioselective SFC include [184] (SFC moving bed technology; 8 refs), [185] (pharmaceutical applications; 32 refs), [186] (different classes of chiral stationary phases in SFC; comparisons of SFC with other chromatographic techniques; 57 refs), [187] (separation of polar analytes by packed column SFC; 221 refs), and [188] (enantiomer separation by packed column SFC; selector-analyte interactions involved in the enantiodiscrimination process; 72 refs).

2.6
Other Methods for the Elucidation of Molecular Structures and Mechanistic Details of Chiral Pollutants

As outlined in Section 2.3.4, the lack of standards of very high optical purity may be a drawback for reliable peak assignments and quantitations. Furthermore, enantioselective analyses are increasingly part of toxicological and mechanistic studies that aim at enzyme/pollutant interactions and deepened insight into the effects of chiral drugs [133, 195]. Therefore, the preparation or separation as well as the determination of the absolute configuration of enantiopure compounds is a crucial part of such endeavours. Very powerful methods that can be employed for structure elucidations and/or mechanistic studies include X-ray crystallography and nuclear magnetic resonance (NMR) spectroscopy.

2.6.1
X-ray Crystallography

The condition for the application of this experimental approach is the separation of a pure enantiomer, which may have been achieved by separating a sufficient amount of the enantiopure compound by enantioselective HPLC [152, 196] or preparative cGC [197]. As an example of this experimental approach, the determination of the absolute structure of (+)-α-HCH is given [152, 196]: the HPLC separation of the (+)-α-HCH enantiomer was carried out on a LiChrocart® 250-4 ChiraDex (5 μm) column (Merck, Darmstadt). Recrystallisation from diethyl ether afforded single crystals of this enantiomer which were suitable for X-ray analysis. The exact parameters used for the X-ray analysis can be found in the original papers [152, 196]. The molecular structure of (+)-α-1S,2R,3R,4S,5S,6S-hexachlorocyclohexane thus determined is shown in Fig. 2.5.

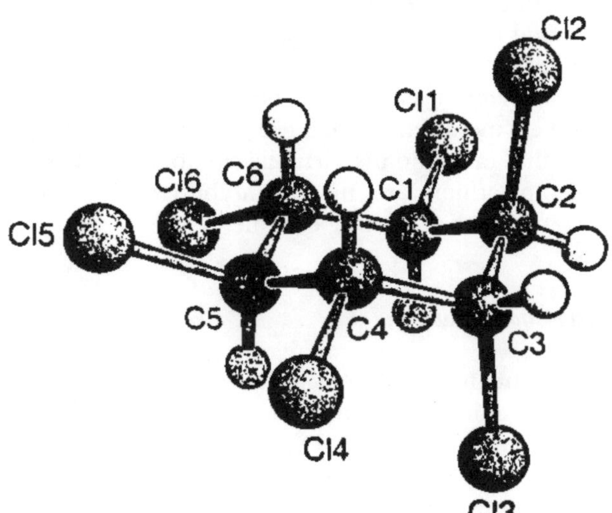

Fig. 2.5. Molecular structure of (+)-α-1S,2R,3R,4S,5S,6S-hexachlorocyclohexane. Bond lengths (pm): Cl1–C1 178.3; Cl2–C2 179.3; Cl3–C3 178.5; Cl4–C4 179.3; Cl5–C5 178.5; Cl6–C6 178.8; C1–C2 152.3; C2–C3 151.6; C3–C4 152.8; C4–C5 152.1; C5–C6 153.1; C6–C1 151.7; all C–H bond lengths are 98.0 pm; selected bond angles (°): Cl1–C1–C2 110.0; Cl2–C2–C3 108.0; Cl3–C3–C4 112.0; Cl4–C4–C5 109.9; Cl5–C5–C6 109.8; Cl6–C6–C1 109.9 [196]

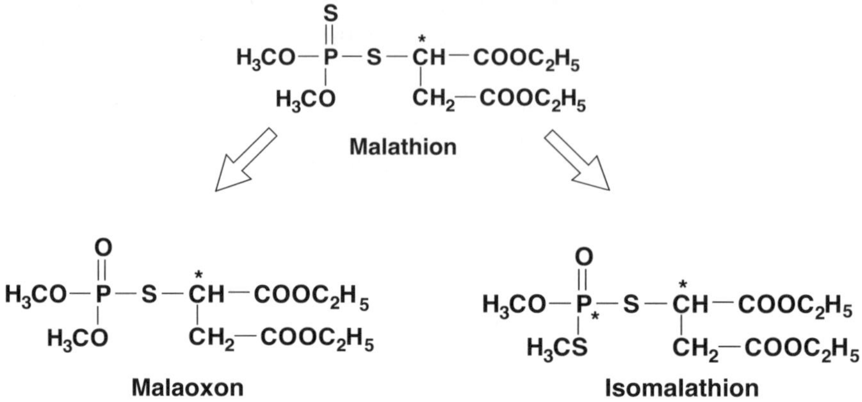

Fig. 2.6. Molecular structures of malathion, malaoxon and isomalathion

Another example is the X-ray crystallographic determination of the absolute configuration of the four stereoisomers of isomalathion by analysis of an alkaloid salt precursor [198, 199]. Again, all four stereoisomers were chromatographically resolved by enantioselective HPLC, in this case using the Chiralpak AD chiral stationary phase. Isomalathion results from the thermal isomerisation of malathion, one of the most widely used organophosphorus insecticides, and has been identified in certain commercial formulations. In this transformation to both malaoxon and isomalathion, the stereogenic centre present initially on the succinyl ligand of malathion is maintained (Fig. 2.6). Further, the isomerisation to isomalathion forms an asymmetric phosphorus atom, leading to the four possible stereoisomers mentioned above. From a toxicological point of view it is important to note that malathion is virtually nontoxic, while the racemates of malaoxon and isomalathion are potent inhibitors of acetylcholinesterase (AChE), a neurotransmitter-mediating enzyme.

2.6.2
Nuclear Magnetic Resonance Studies

Nuclear magnetic resonance (NMR) techniques, employing appropriate chiral shift reagents such as chiral europium(III) or praseodym(III) derivatives (see chapter 5.2.1), represent a powerful tool that can be considered an alternative to X-ray analysis with regard to the determination of the absolute structures of chiral environmental pollutants. Examples of applications to synthesised or separated enantiopure standard compounds of malathion, malaoxon and isomalathion enantiomers and diastereomers (see Fig. 2.6), respectively, have been published by Berkman et al. [198, 199]. Fingerling and Parlar [201] identified the structure of a cyclic ketone, formed from the toxaphene components Toxicant A (Parlar42) and Toxicant B (Parlar32) in a flooded loamy silt (i.e., anaerobic conditions) in laboratory experiments, as 7*b*,8*c*,9*c*-trichlorocamphen-2-one using 400 MHz proton NMR spectroscopy (DQF-COSY and NOE difference spectra).

Furthermore, important mechanistic details about molecular interactions may be obtained from NMR studies [133]. Investigations with unmodified cyclodextrins showed that they may function as chiral shift reagents in solution together with a racemic substrate [173]. For the S-enantiomer of 1-phenyl-2,2,2-trifluoroethanol, the CF$_3$-resonance in the ^{19}F-NMR spectrum is shifted by 4.3 Hz to a higher field in the presence of α-cyclodextrin. This effect is attributed to selective inclusion of the enantiomer to the chiral host α-cyclodextrin. In the same work chemical nonequivalence between enantiotopic prochiral CF$_3$ groups induced by β-cyclodextrin was observed. In contrast to chiral lanthanide shift reagents, cyclodextrins cause nonequivalence even in the absence of functional groups. Further examples, including a detailed discussion on correlations between chemical shift nonequivalence and host-guest interactions between racemic compounds and cyclodextrins, can be found in the monograph by König [133].

References

1. König WA (1987) The practice of enantiomer separation by capillary gas chromatography. Hüthig, Heidelberg, p 168
2. Jaus A, Oehme M (1999) Enantioselective behaviour of ethylated γ-cyclodextrins as GC stationary phases for chlorinated pesticides and phase characterisation by HPLC. Chromatographia 50:299–304
3. Vetter W, Klobes U, Luckas B, Hottinger G (1997) Enantiomeric resolution of persistent compounds of technical toxaphene (CTTs) on t-butyldimethylsilylated β-cyclodextrin phases. Chromatographia 45:255–262
4. Oehme M, personal communication
5. Vetter W, Klobes U, Luckas B, Hottinger G (1998) Enantioselective determination of persistent toxaphene compounds: possibilities on and alternatives to $tert$-butyldimethylsilylated β-cyclodextrin. Organohalogen Compd 35:305–308
6. Koizumi K, Kubota Y, Utamura T, Horiyama (1986) Analysis of heptakis(2,6-di-O-methyl)-β-cyclodextrin by thin-layer, high-performance liquid and gas chromatography and mass spectrometry. J Chromatogr 368:329–337
7. Jaus A, Oehme M (1999) Benefits of partially alkylated cyclodextrins for the enantioselective separation of chiral polychlorinated compounds. Organohalogen Compd 40:387–390
8. Grob K Jr, Grob G, Grob K (1978) Comprehensive, standardized quality test for glass capillary columns. J Chromatogr 156:1–20
9. Grob K Jr, Grob G, Grob K (1981) Testing capillary gas chromatography columns. J Chromatogr 219:13–20
10. Taylor DR, Maher K (1992) Chiral separations by high-performance liquid chromatography. J Chromatogr Sci 30:67–85
11. Armstrong DW (1984) Chiral stationary phases for high performance liquid chromatographic separation of enantiomers: a mini-review. J Liq Chromatogr 7:353–376
12. Armstrong DW (1987) Optical isomer separation by liquid chromatography. Anal Chem 59:84A-91A
13. Armstrong DW, Han SM (1988) Enantiomeric separations in chromatography. CRC Crit Rev Anal Chem 19:175–224
14. Krstulovic AM (ed) (1989) Chiral separations by HPLC. Ellis Horwood, Chichester, UK
15. Allenmark S (1991) Chromatographic enantioseparation, methods and applications, Ellis Horwood, Chichester, UK, p 282

16. Ahuja S (ed) (1997) Chiral separations: applications and technology, ACS, Washington DC, p 349
17. Armstrong DW (1998) The evolution of chiral stationary phases for liquid chromatography. LC-GC Intl 22–31
18. Dalgliesh CE (1952) Optical resolution of aromatic amino acids on paper chromatograms. J Chem Soc 47:3940–3942
19. Easson LH, Stedman E (1933) Studies on the relationship between chemical constitution and physiological action. V. Molecular dissymmetry and physiological activity. Biochem J 27:1257–1266
20. Kotake M, Saken T, Nakamura N, Senoh S (1951) Resolution of enantiomers of some amino acids by paper chromatography. J Am Chem Soc 73:2973–2974
21. Karagounis G, Coumoulos (1938) A new method of resolving a racemic compound. Nature 142:162–163
22. Henderson GM, Rule HG (1938) Resolving a racemic compound. Nature 141:917–919
23. Krebs H, Rasche R (1954) Über die chromatographische Spaltung von Racematen I. Optisch aktive Kobaltkomplexe von Dithiosäuren. Z Anorg Chem 276:236–238
24. Krebs H, Schümacher W (1966) Über die chromatographische Spaltung von Racematen. IV. Camphersäuren, Diphensäuren und Aminosäuren. Chem Ber 99:1341–1346
25. Steckelberg W, Bloch M, Musso HJ (1968) Notiz zur Antipodentrennung von Biphenylderivaten durch Chromatographie. Chem Ber 101:1519–1521
26. Taylor LT, Busch DH (1967) Chromatographic resolution of the antipodes of a helical complex of nickel(II). J Am Chem Soc 89:5372–5376
27. Leitch RE, Rothbart HL, Rieman W III (1967) Partial resolution of mandelic acid with Sephadex gels. J Chromatogr 28:132–136
28. Karagounis G, Charbonnier E, Floss E (1959) Chromatographic resolution of racemic compounds. J Chromatogr 2:84–89
29. Klemm LH, Reed D (1960) Optical resolution by molecular complexation chromatography. J Chromatogr 3:364–368
30. Grubhofer N, Schleith L (1953) Modifizierte Ionenaustauscher als spezifische Adsorbentien. Naturwissenschaften 40:508–512
31. Manecke G, Lamer W (1967) Zur Racemattrennung an optisch aktiven Polymeren. Naturwissenschaften 54:140–142
32. Blaschke G (1971) Chromatographic resolution of racemates. Angew Chem Int Ed Engl 10:520–521
33. Blaschke G (1974) Chromatographie racemischer Mandelsäure an Polyacrylsäure-estern und -amiden optisch aktiver Ephedrinderivate. Chem Ber 107:237–252
34. Buss DR, Vermeulan T (1968) Optical isomer separation. Ind Eng Chem 60:12–29
35. Davankov VA (1980) Resolution of racemates by ligand exchange chromatography. Adv Chromatogr 18:139–195
36. Wainer IW (1988) A practical guide to the selection and use of HPLC chiral stationary phases. J.T. Baker, Phillipsburg, NJ, USA
37. Pirkle WH, Pochapsky TC (1989) Considerations of chiral recognition relevant to the liquid chromatographic separation of enantiomers. Chem Rev 89:347–362
38. Doury-Berthod M, Poitrenaud C, Tremillon B (1977) Ligand-exchange separation of amino acids. I. Distribution equilibria of some amino acids between ammoniacal copper(II) nitrate solution and phosphonic, carboxylic, and iminodiacetic ion exchangers in the copper(II) form. J Chromatogr 131:73–77
39. Davankov VA, Semechkin AV (1977) Ligand-exchange chromatography. J Chromatogr 141:313–353
40. Mikes F, Boshart G, Gil-Av E (1976) Resolution of optical isomers by high-performance liquid chromatography, using coated and bonded chiral charge-transfer complexing agents as stationary phases. J Chromatogr 122:205–221
41. Matlin SA, Stacey VE, Lough WJ (1988) Hexahelicene chiral stationary phase. I. Phase synthesis and use in HPLC resolution of enantiomers. J Chromatogr 450:157–163

42. Liu RH, Ku WW (1983) Chiral stationary phases for the gas-liquid chromatographic separation of enantiomers. J Chromatogr 271:309–323

43. Lough WJ (ed) (1989) Chiral liquid chromatography. Blackie & Sons, Glasgow, Scotland

44. Dappen R, Arm H, Meyer VR (1986) Applications and limitations of commercially available chiral stationary phases for HPLC. J Chromatogr 373:1–20

45. Pirkle WH, Pochapsky TC (1986) Generation of extreme selectivity in chiral recognition. J Chromatogr 369:175–177

46. Pirkle WH, Finn JM, Schreiner JL, Hamper BC (1981) A widely used chiral stationary phase for the high-performance liquid chromatographic separation of enantiomers. J Am Chem Soc 103:3964–3966

47. Doyle TD, Adams WM, Fry FS, Wainer IW (1986) The application of HPLC chiral stationary phases to stereochemical problems of pharmaceutical interest: a general method for the resolution of enantiomeric amines as α-naphthylcarbamate derivatives. J Liq Chromatogr 9:455–471

48. Wainer IW, Doyle TD (1984) Application of high-performance liquid chromatographic chiral stationary phases to pharmaceutical analysis: structural and conformational effects in the direct enantiomeric resolution of α-methylarylacetic antiinflammatory agents. J Chromatogr 284:117–124

49. Dotsevi G, Sogah Y, Cram DJ (1975) Chromatographic optical resolution through chiral complexation of amino ester salts by a host covalently bound to silica gel. J Am Chem Soc 97:1259–1261

50. Udvarhelyi PM, Sunter DC, Watkins JC (1990) Direct separation of amino acid enantiomers using a chiral crown ether stationary phase. Application to 2-amino-ω-phosphonoalkanoic acids. J Chromatogr 519:69–74

51. Armstrong DW, DeMond W (1984) Cyclodextrin bonded phases for the liquid chromatographic separation of optical, geometrical, and structural isomers. J Chromatogr Sci 22:411–421

52. Armstrong DW (1985) Bonded phase material for chromatographic separations. US Patent 4,539,399

53. Hesse G, Hagel R (1973) A complete separation of a racemic mixture by elution chromatography on cellulose triacetate. Chromatographia 6:277–278

54. Lindner KR, Mannschreck A (1980) Separation of enantiomers by HPLC on triacetylcellulose. J Chromatogr 193:308–310

55. Köller H, Rimbock K-H, Mannschreck A (1983) Characterization of a sorbent for the separation of enantiomers. J Chromatogr 282:89–95

56. Mintas M, Mannschreck A, Schneider MP (1997) (1S,2S)-(+)- and (1R,2R)-(–)-1,2-Diphenylcyclopropane from their racemic mixture by liquid chromatography on triacetylcellulose. J Chem Soc Chem Commun 602–603

57. Rizzi AM (1990) Coupled column chromatography in chiral separations. I. Enantiomeric separations on swollen microcrystalline cellulose triacetate columns after a preseparation on a non-chiral alkylsilica column. J Chromatogr 513:192–207

58. Okamoto Y, Kawashima M, Aburatani R, Hatada K, Nishiyama T, Masuda M (1986) Optical resolution of β-blockers by HPLC on cellulose triphenylcarbamate derivatives. Chem Lett 1237–1243

59. Allenmark S, Bomgren B, Boren H (1982) Direct resolution of enantiomers by liquid affinity chromatography on albumin-agarose under isocratic conditions. J Chromatogr 237:473–481

60. Allenmark S, Bomgren B (1982) Direct liquid chromatographic separation of enantiomers on immobilized protein stationary phases. II. Optical resolution of a sulfoxide, a sulfoxamine, and a benzoylamino acid. J Chromatogr 25:297–303

61. Hermansson J, Eriksson M (1986) Direct liquid chromatographic resolution of acidic drugs using a chiral α_1-acid glycoprotein column. J Liq Chromatogr 9:621–639

62. Krstulovic AM, Vende JL (1989) Improved performance of the second generation α_1-AGP columns: applications to the routine assay of plasma levels of alfuzosin hydrochloride. Chirality 1:243–245
63. Ôi N, Kitahara H, Doi T (1988) European Patent EP029703, 15 July 1988
64. Pirkle WH, Welch CJ (1992) An improved chiral stationary phase for the chromatographic separation of underivatized naproxen enantiomers. J Liq Chromatogr 15:1947–1955
65. Armstrong DW, Hilton M, Coffin L (1992) Multimodal chiral stationary phases for liquid chromatography: (R)- and (S)-naphthylethylcarbamate-derivatized β-cyclodextrin. LC-GC Intl 5:28–36
66. (1996) Cyclobond handbook. Advanced Separation Technologies Inc., Whippany, NJ, USA
67. Stalcup AM, Chang S-C, Armstrong DW (1990) (S)-2-Hydroxypropyl-β-cyclodextrin, a new chiral stationary phase for reversed-phase liquid chromatography. J Chromatogr 513:181–194
68. Chang SC, Wang LR, Armstrong DW (1992) Facile resolution of N-tert-butoxycarbonyl amino acids: the importance of enantiomeric purity in peptide synthesis. J Liq Chromatogr 15:1411–1429
69. Zukowski J, Pawlowska M, Armstrong DW (1992) Efficient enantioselective separation and determination of trace impurities in secondary acids (i.e., imino acids). J Chromatogr 623:33–41
70. Zukowski J, Pawlowska M, Nagatkina M, Armstrong DW (1993) High-performance liquid chromatographic enantioseparation of glycyl di- and tripeptides on native cyclodextrin bonded phases. Mechanistic considerations. J Chromatogr 629:169–179
71. Pawlowska M, Chen S, Armstrong DW (1993) Enantiomeric separation of fluorescent, 6-aminoquinolyl-N-hydroxysuccinimidyl carbamate, tagged amino acids. J Chromatogr 641:257–265
72. Armstrong DW, Chen S, Chang C, Chang S (1992) A new approach for the direct resolution of racemic beta adrenergic blocking agents by HPLC. J Liq Chromatogr 15:545–556
73. Chang SC, Reid GL III, Chen S, Chang CD, Armstrong DW (1993) Evaluation of a new polar-organic high-performance liquid chromatographic mobile phase for cyclodextrin-bonded chiral stationary phases. Trends Anal Chem 12:144–153
74. Armstrong DW, Tang Y, Chen S, Zhou Y, Bagwill C, Chen J-R (1994) Macrocyclic antibiotics as a new class of chiral selectors for liquid chromatography. Anal Chem 66:1473–1484
75. Chen S, Liu Y, Armstrong DW, Borrell JI, Martinez-Teipel B, Matallana J L (1995) Enantioresolution of substituted 2-methoxy-6-oxo-1,4,5,6-tetrahydropyridine-3-carbonitriles on macrocyclic antibiotic and cyclodextrin stationary phases. J Liq Chromatogr 18:1495–1507
76. Armstrong DW, Liu Y, Ekborg-Ott KH (1995) A covalently bonded teicoplanin chiral stationary phase for HPLC. Chirality 7:474–497
77. Berthod A, Liu Y, Bagwill C, Armstrong DW (1996) Facile LC enantioresolution of native amino acids and peptides using teicoplanin chiral stationary phase. J Chromatogr A 731:123–137
78. (1996) Chirobiotic handbook. Advanced Separation Technologies Inc., Whippany, NJ, USA
79. Ringo MC, Evans CE (1998) Liquid chromatography as a measurement tool for chiral interactions. Anal Chem News & Features 315A-321A
80. Locke DC (1976) In: Advances in chromatography, vol 14. Giggings JC, Grushka E, Cazes J, Brown PR (eds) Marcel Dekker, New York, 1976, pp 87–198
81. de Ligny CL (1976) In: Giggings JC, Grushka E, Cazes J, Brown PR (eds) Advances in chromatography, vol 14. Marcel Dekker, New York, pp 265–304
82. Hafkenscheid TL, Tomlinson E In: Giggings JC, Grushka E, Cazes J, Brown PR, (eds) Advances in chromatography, vol 25. Marcel Dekker, New York, pp 2–62

83. Giddings JC (1965) Dynamics of chromatography Part 1: Principles and theory. Marcel Dekker, New York

84. Noctor TAG, Felix G, Wainer IW (1991) Stereochemical resolution of enantiomeric 2-arylpropionic acid non-steroidal anti-inflammatory drugs on a human serum albumin based high-performance liquid chromatographic chiral stationary phase. Chromatographia 31:55–59

85. Wainer IW (1994) Enantioselective high-performance liquid affinity chromatography as a probe of ligand-biopolymer interactions: an overview of a different use for high-performance liquid chromatographic chiral stationary phases. J Chromatogr 666:221–234

86. Domenici E, Bertucci C, Salvadori P, Motellier S, Wainer IW (1990) Immobilized serum albumin: rapid HPLC probe of stereoselective protein-binding interactions. Chirality 2:263–268

87. Cancelliere G, DÁcquarica L, Gasparrini F, Misiti D, Villani C (1999) Enantioselective chromatographic techniques. Principles and state of the art. Chim Ind (Milan) 81:475–480

88. Welch CJ, Protopopova MN, Bhat GA (1999) Microscale synthesis and screening of combinatorial libraries of new chromatographic stationary phases. Spec Publ – R Soc Chem 235:129–138

89. Bojarski J, Aboul-Enein HY (1999) Recent chromatographic and electrophoretic enantioseparations of cardiovascular drugs. Biomed Chromatogr 13:197–208

90. Tanaka N (1999) An approach to high-performance packing materials for HPLC. Chromatography 20:1–10

91. Hage DS (1999) Affinity chromatography: a review of clinical applications. Clin Chem 45:593–615

92. Ishida K, Matsuda H, Murakami M, Yamaguchi K (1999) Biologically active compounds of the cyanobacterium Microcystis aeruginosa. Tennen Yuki Kagobutsu Toronkai Koen Yoshishu 1997, 39:667–672

93. Makino S (1999) Scaling-up procedures on chiral separation by HPLC. Yuki Gosei Kagaku Kyokaishi 57:472–477

94. Hühnerfuss H, Kallenborn R (1992) Review. Chromatographic separation of marine organic pollutants. J Chromatogr 580:191–214

95. Hühnerfuss H, Faller J, Kallenborn R, König WA, Ludwig P, Pfaffenberger B, Oehme M, Rimkus G (1993) Enantioselective and nonenantioselective degradation of organic pollutants in the marine ecosystem. Chirality 5:393–399

96. Vetter W, Schurig V (1997) Enantioselective determination of chiral organochlorine compounds in biota by gas chromatography on modified cyclodextrins. J Chromatogr A774:143–175

97. Hühnerfuss H (1992) Chromatographische Trennung von Enantiomeren organischer Schadstoffe – ein neuer Zugang zu Prozeßstudien im marinen Ökosystem. GIT Fachz Lab 36:489–497

98. Hühnerfuss H (1998) Chromatographic enantiomer separation of chiral xenobiotics and their metabolites – a versatile tool for process studies in marine and terrestrial ecosystems. Organohalogen Compd 35:319–324

99. Hühnerfuss H (2000) Chromatographic enantiomer separation of chiral xenobiotics and their metabolites – a versatile tool for process studies in marine and terrestrial ecosystems. Chemosphere 40:913–919

100. Gil-Av E, Feibush B, Charles-Sigler R (1966) In: Littlewood AB (ed) Gas chromatography. Institute of Petroleum, London, p 227

101. Gil-Av E, Feibush B, Charles-Sigler R (1966) Separation of enantiomers by gas liquid chromatography with an optically active stationary phase. Tetrahedron Lett 1009–1015

102. König WA, Parr W, Lichtenstein HA, Bayer E, Oró J (1970) Gas chromatographic separation of amino acids and their enantiomers: non-polar stationary phases and a new optically active phase. J Chromatogr Sci 8:183–186

103. König WA, Nicholson GJ (1975) Glass capillaries for fast gas chromatographic separation of amino acid enantiomers. Anal Chem 47:951–952

104. Feibush B, Gil-Av E (1967) Gas chromatography with optically active stationary phases. Resolution of primary amines. J Gas Chromatogr 5:257–260
105. Feibush B, Gil-Av E, Tamari T (1972) Assignment of the configuration of optical isomers by gas chromatography with asymmetric phases. Order of emergence of aminoalkanes, and α-, β-, and γ-amino acids on carbonylbis(N-L-valine isopropyl ester) J Chem Soc Perkin 2 1197–1203
106. Feibush B (1971) Interaction between asymmetric solutes and solvents. N-Lauroyl-L-valyl-t-butylamide as stationary phase in gas liquid partition chromatography. Chem Commun 544–545
107. Weinstein S, Feibush B, Gil-Av E (1976) N-Acyl derivatives of chiral amines as novel, readily prepared phases for the separation of optical isomers by gas chromatography. J Chromatogr 126:97–111
108. König WA, Sievers S, Schulze U (1980) Enantiomerentrennung von 2-Hydroxycarbonsäuren an optisch aktiven stationären Phasen. Angew Chem 92:935–936; Angew Chem Int Ed Engl 19:910
109. König WA, Sievers S (1980) Structural requirements for enantioselectivity in gas chromatography of chiral α-hydroxy acids. J Chromatogr 200:189–194
110. König WA, Benecke I, Sievers S (1981) New results in the gas chromatographic separation of enantiomers of hydroxy acids and carbohydrates. J Chromatogr 217:71–79
111. Ôi N, Doi T, Kitahara H, Inda Y (1982) Gas chromatographic determination of optical isomers of some carboxylic acids and amines with optically active stationary phases. J Chromatogr 239:493–498
112. Ôi N, Kitahara H, Doi T (1983) Gas chromatographic separation of chrysanthemic acid ester enantiomers on a novel chiral stationary phase. J Chromatogr 254:282–284
113. Ôi N, Doi T, Kitahara H, Inda Y (1981) Direct separation of some alcohol enantiomers by gas chromatography with amino acid derivatives as chiral stationary phases. J Chromatogr 208:404–408
114. Frank H, Nicholson GJ, Bayer E (1977) Rapid gas chromatographic separation of amino acid enantiomers with a novel chiral stationary phase. J Chromatogr Sci 15:174–176
115. Frank H, Nicholson GJ, Bayer E (1978) Chirale Polysiloxane zur Trennung von optischen Antipoden. Angew Chem 90:396–398; Angew Chem Int Ed Engl 17:363
116. Schurig V, Nowotny H-P (1990) Gaschromatographische Enantiomerentrennung an Cyclodextrinderivaten. Angew Chem 102:969–986; Angew Chem Int Ed Engl 29:939
117. Bayer E (1983) Chirale Erkennung von Naturstoffen an optisch aktiven Polysiloxanen. Z Naturforsch 38b:1281–1291
118. Schurig V (1984) Gaschromatographische Enantiomerentrennung an metallkomplexfreien Stationärphasen. Angew Chem 96:733–752; Angew Chem Int Ed Engl 23:747
119. Koppenhoefer B, Bayer E (1985) In: Proceedings of the AJP Martin Honorary Symposium, Urbino, p 1; J Chromatogr Libr 32:1
120. Saeed T, Sandra P, Verzele M (1979) Synthesis and properties of a novel chiral stationary phase for the resolution of amino acid enantiomers. J Chromatogr 186:611–618
121. Saeed T, Sandra P, Verzele M (1980) GC separation of the enantiomers of proline and secondary amines. J High Res Chromatogr Chromatogr Commun 3:35–36
122. König WA, Benecke I (1981) Gas chromatographic separation of enantiomers of amines and amino alcohols on chiral stationary phases. J Chromatogr 209:91–95
123. König WA, Sievers S, Benecke I (1981) In: Kaiser RE (ed) Proceed IVth Intern Symp Capillary Chromatogr. Institut für Chromatographie, Bad Dürkheim and A Hüthig Verlag, Heidelberg, p 703
124. König WA (1982) Separation of enantiomers by capillary gas chromatography with chiral stationary phases. J High Res Chromatogr Chromatogr Commun 5:588–595
125. König WA (1984) In: Schreier P (ed) Analysis of volatiles, de Gruyter, Berlin, p 77
126. Chrompack International, Middelburg, Netherlands

127. Schomburg G, Benecke I, Severin G (1985) Modified GC-phases for the enantiomeric separation of optically active compounds: crosslinking experiments with polymeric chiral compounds. J High Res Chromatogr Chromatogr Commun 8:391–394

128. Liu RH, Ku WW (1983) Chiral stationary phases for the gas-liquid chromatographic separation of enantiomers. J Chromatogr 271:309–323

129. Schurig V (1977) Enantiomerentrennung eines chiralen Olefins durch Komplexierungschromatographie an einem optisch aktiven Rhodium(I)-Komplex. Angew Chem 89:113–114; Angew Chem Int Ed Engl 16:1

130. Schurig V, Bürkle W (1982) Extending the scope of enantiomer resolution by complexation gas chromatography. J Am Chem Soc 104:7573–7580

131. Koscielski T, Sybilska D, Jurczak J (1983) Separation of α- and β-pinene into enantiomers in gas-liquid chromatography systems via α-cyclodextrin inclusion complexes. J Chromatogr 280:131–134

132. Schurig V (1994) Review. Enantiomer separation by gas chromatography on chiral stationary phases. J Chromatogr 666:111–129

133. König WA (1992) Gas chromatographic enantiomer separation with modified cyclodextrins. Hüthig Buch Verlag, Heidelberg

134. Juvancz Z, Alexander G, Szejtli J (1987) Permethylated β-cyclodextrin as stationary phase in capillary gas chromatography. J High Resolut Chromatogr Chromatogr Commun 10:105–107

135. Alexander G, Juvancz Z, Szejtli J (1988) Cyclodextrins and their derivatives as stationary phases in GC capillary columns. J High Resolut Chromatogr Chromatogr Commun 11:110–113

136. Venema A, Tolsma PJA (1989) Enantiomer separation with capillary gas chromatography columns coated with cyclodextrins. I. Separation of enantiomeric 2-substituted propionic acid esters and some lower alcohols with permethylated β-cyclodextrin. J High Resolut Chromatogr 12:32–34

137. Schurig V, Nowotny H-P (1988) Separation of enantiomers on diluted permethylated β-cyclodextrin by high-resolution gas chromatography. J Chromatogr 441:155–163

138. König WA, Lutz S (1988) In: Holmstedt B, Frank H, Testa B (eds) Chirality and biological activity. Proc Intern Symp, April 5–8, Tübingen; (1990) AR Liss Inc., New York, p 55

139. König WA, Lutz S, Wenz G (1988) Modifizierte Cyclodextrine – neue, hochenantioselektive Trennphasen für die Gaschromatographie. Angew Chem 100:989–90; Angew Chem Int Ed Engl 27:979–980

140. Schomburg G, Weeke F, Müller F, Oreans M (1982) Multidimensional gas chromatography (MDC) in capillary columns using double oven instruments and a newly designed coupling piece for monitoring detection after pre-separation. Chromatographia 16:87–91

141. Duinker JC, Schulz DE, Petrick G (1988) Multidimensional gas chromatography with electron capture detection for the determination of toxic congeners in polychlorinated biphenyl mixtures. Anal Chem 60:478–482

142. Schulz DE, Petrick G, Duinker JC (1989) Complete characterization of polychlorinated biphenyl congeners in commercial Aroclor and Clophen mixtures by multidimensional gas chromatography-electron capture detection. Environ Sci Technol 23:852–859

143. Saenger W (1980) Cyclodextrin-Einschlußverbindungen in Forschung und Industrie. Angew Chem 92:343–361

144. Schurig V, Glausch A (1993) Enantiomer separation of atropisomeric polychlorinated biphenyls (PCBs) by gas chromatography on Chirasil-Dex. Naturwissenschaften 80:468–469

145. Glausch A, Nicholson GJ, Fluck M, Schurig V (1994) Separation of the enantiomers of stable atropisomeric polychlorinated biphenyls (PCBs) by multidimensional gas chromatography on Chirasil-Dex. J High Resolut Chromatogr 17:347–349

146. Glausch A, Hahn J, Schurig V (1995) Enantioselective determination of chiral 2,2',3,3',4,6'-hexachlorobiphenyl (PCB 132) in human milk samples by multidimensional gas chromatography/electron capture detection and by mass spectrometry. Chemosphere 30:2079–2085

147. Blanch GP, Glausch A, Schurig V, Serrano R, González MJ (1996) Quantification and determination of enantiomeric ratios of chiral PCB 95, PCB 132, and PCB 149 in shark liver samples (*C. coelolepis*) from the Atlantic ocean. J High Resolut Chromatogr 19:392–396

148. Koske G, Leupold G, Angerhöfer D, Parlar H (1998) Multidimensional gas chromatographic enantiomer quantification of some polycyclic xenobiotics in cod liver and fish oils. Organohalogen Compd 35:363–366

149. Reich S, Jiménez, Marsili L, Hernández LM, Schurig V, González MJ (1998) Enantiomeric ratios of chiral PCBs in striped dolphins (*Stenella coeruleoalba*) from the Mediterranean Sea. Organohalogen Compd 35:335–338

150. Hinze WL, Williams RW Jr, Fu ZS, Suzuki Y, Quina FH (1990) Novel chiral separation techniques based on surfactants. Colloids Surf 48:79–94

151. Armstrong DW, Jin HL (1987) Enrichment of enantiomers and other isomers with aqueous liquid membranes containing cyclodextrin carriers. Anal Chem 59:2237–2241

152. Möller K (1993) Untersuchungen zur enantioselektiven Anreicherung von chiralen Schadstoffen im marinen und terrestrischen Ökosystem. Master Thesis, University of Hamburg, Germany p 49

153. Oehme M, Kallenborn R, Wiberg K, Rappe C (1994) Simultaneous enantioselective separation of chlordanes, a nonachlor compound, and *o,p*-DDT in environmental samples using tandem capillary columns. J High Resolut Chromatogr 17:583–588

154. Jorgenson JW, Lukacs KD (1981) Zone electrophoresis in open tubular glass capillaries. Anal Chem 53:1298–1302

155. Terabe S, Otsuka K, Ando T (1985) Electrokinetic chromatography with micellar solution and open-tubular capillary. Anal Chem 57:834–841

156. Chankvetadze B, Frost M, Blaschke G (1999) Kapillarelektrophorese, eine attraktive Methode zur Enantiomerenanalytik. Pharm u Z 28:186–196

157. Nishi H (1996) Review. Enantiomer separation of drugs by electrokinetic chromatography. J Chromatogr A 735:57–76

158. Jakubetz H, Juza M, Schurig V (1998) Duale Enantiomerendiskriminierung unter gleichzeitiger Verwendung zweier β-Cyclodextrinderivate in mobiler und stationärer Phase. GIT 5/98:479–481

159. Majors RE (1998) Analytical HPLC column technology – the current status. LC-GC Intl 7–21

160. Zhong Y, Lin B (1998) Report on 2nd Asia-Pacific International Capillary Electrophoresis and Related Microscale Techniques (APCE'98), Dalian, China, October 8–11

161. Benito I, Marina ML, Diez-Masa, Gonzalez MJ (1996) Separation of chiral polychlorinated biphenyls by cyclodextrin modified micellar electrokinetic chromatography. Organohalogen Compd 27:323–326

162. Garrison AW, Nzengung VA, Avants JK, Ellington J, Wolfe NL (1997) Determining the environmental enantioselectivity of *o,p'*-DDT and *o,p'*-DDD. Organohalogen Compd 31:256–261

163. Grainger J, Smith P, Smith C, Otsuka K, Lovinggood J, Patterson DG Jr (1998) Chiral separation of *ortho*-substituted polychlorinated biphenyl enantiomers and phenoxy herbicides by capillary electrophoresis with UV and MS detectors. Organohalogen Compd 35:351–354

164. Schurig V, Wistuba D (1999) Recent innovations in enantiomer separation by electrochromatography utilizing modified cyclodextrins as stationary phases. Electrophoresis 20:2313–2328

165. Okamoto M, Okumura T (1999) On-line CE-MS – present status and future prospect. Chromatography 20:19–26

166. Chankvetadze B (1999) Recent trends in enantioseparations using capillary electromigration techniques. Trends Anal Chem 18:485–498
167. Hui F, Caude M (1999) Enantioseparations in CE using macrocyclic antibiotics as chiral selectors. Analusis 27:131–137
168. Zhu L, Xu X, Lin B (1999) Temperature effect and temperature gradient technology in capillary electrophoresis. Sepu 17:21–25
169. Waetzig H, Degenhardt M, Kunkel A (1998) Strategies for capillary electrophoresis. Method development and validation for pharmaceutical and biological applications. Electrophoresis 19:2695–2752
170. Hutt AJ, Patel BK (1998) Enantiospecific bioanalysis: techniques and applications. Biomed Health Res 25:196–212
171. Schweitz L, Andersson LI, Nilsson S (1998) Molecular imprint-based stationary phases for capillary electrochromatography. J Chromatogr A 817:5–13
172. Castellan GW (1995) Physical chemistry. Benjamin Cummings, Reading, MA, USA, p 942
173. Altria KD, Kelly MA, Clark BJ (1998) Current applications in the analysis of pharmaceuticals by capillary electrophoresis. II. Trends Anal Chem 17:214–226
174. Khaledi MG (1998) High-performance capillary electrophoresis: theory, techniques and applications. Chemical Analysis, vol 146. Wiley
175. Landers JP (1997) Handbook of capillary electrophoresis. CRC Press
176. Chankvetadze B (1997) Electrophoresis in chiral analysis. Wiley, Chichester
177. Ward TJ (1994) Chiral media for capillary electrophoresis. Anal Chem 66:632–640A
178. Novotny M, Soini H, Steffansson M (1994) Chiral separation through capillary electromigration methods. Anal Chem 66:646–665A
179. Bartle KD (1988) In: Smith RM (ed) Supercritical fluid chromatography. The Royal Society of Chemistry, London
180. Schurig V, Ossig A, Link R (1989) Temperaturabhängige Umkehr der Enantioselektivität bei der Komplexierungs-Gaschromatographie an chiralen Phasen. Angew Chem 101:197–198
181. Schleimer M, Schurig V (1992) In: Wenclawiak B (ed) Analysis with supercritical fluids. Springer, Berlin, p 134
182. Schurig V, Schleimer M, Jung M (1992) Separation of enantiomers by capillary SFC on Chirasil-Dex and Chiralsil-Metal. Proc 3rd Int Symp Chiral Dicrimination, Tübingen, October 5–8, p 76
183. Jung M, Mayer S, Schurig V (1994) Enantiomer separation by GC, SFC, and CE on immobilized polysiloxane-bonded cyclodextrins. LC-GC Int 7:340–347
184. Juza M, Mazzotti M, Morbidelli M (1998) Simulated moving-bed technology. Analytical separations on a large scale. GIT Spez Chromatogr 18:70, 72–74, 76
185. Maftouh M (1997) Supercritical fluid chromatography: a procedure developed for chiral analysis for pharmaceutical products. Spectra Anal 26:25–28
186. Williams KL, Sander LC (1997) Enantiomer separations on chiral stationary phases in supercritical fluid chromatography. J Chromatogr A 785:149–158
187. Berger TA (1997) Separation of polar solutes by packed column supercritical fluid chromatography. J Chromatogr A 785:3–33
188. Wolf C, Pirkle WH (1997) Enantiomer separation by supercritical fluid chromatography on packed columns. LC-GC 15:352, 354–361, 363
189. Baycan-Keller R, Oehme M (1999) Optimization of tandem columns for the isomer and enantiomer selective separation of toxaphenes. J Chromatogr A 837:201–210
190. Jones JR, Purnell JH (1990) Prediction of retention times in serially linked open-tubular gas chromatographic columns and optimization of column lengths. Anal Chem 62:2300–2306
191. Benicka E, Novakovski R, Hrouzek J, Krupcik J (1996) Multidimensional gas chromatographic separation of selected PCB atropisomers in technical formulations and sediments. J High Resolut Chromatogr 19:95–98

192. de Geus HJ, Baycan-Keller R, Oehme M, de Boer J, Brinkman UATh (1998) Enantiomer ratios of bornane congeners in biological samples using heart-cut gas chromatography with an enantioselective column. J High Resol Chromatogr 21:39–46
193. Schurig V (1988) Enantiomer analysis by complexation gas chromatography. Scope, merits and limitations. J Chromatogr 441:135–153
194. Meyer VR (1995) Accuracy in the chromatographic determination of extreme enantiomeric ratios: a critical review. Chirality 7:567–571
195. Islam MR, Mahdi JG, Bowen JD (1997) Pharmacological importance of stereochemical resolution of enantiomeric drugs. Drug Safety 17:149–165
196. Möller K, Bretzke C, Hühnerfuss H, Kallenborn R, Kinkel JN, Kopf J, Rimkus G (1994) The absolute configuration of $(+)$-α-1,2,3,4,5,6-hexachlorocyclohexane and its permeation through the seal blood-brain barrier. Angew Chem 33:882–884; Angew Chem Int Ed Engl 33:882–884
197. König WA, Hardt IH, Gehrcke B, Hochmuth DH, Hühnerfuss H, Pfaffenberger B, Rimkus G (1994) Optisch aktive Referenzsubstanzen für die Umweltanalytik. Angew Chem 106:2175–2177; Angew Chem Int Ed Engl 33:2085–2087
198. Berkman CE, Thompson CM, Perrin SR (1993) Synthesis, absolute configuration, and analysis of malathion, malaoxon, and isomalathion enantiomers. Chem Res Toxicol 6:718–723
199. Berkman CE, Thompson CM, Perrin SR (1993) Erratum: Synthesis, absolute configuration, and analysis of malathion, malaoxon, and isomalathion enantiomers. Chem Res Toxicol 7:275
200. Nishikawa Y (1993) Enantiomer separation of synthetic pyrethroids by subcritical and supercritical fluid chromatography with chiral stationary phases. Anal Sci 9:33–37
201. Fingerling DM, Parlar H (1997) Spectroscopic characterization of 7b,8c,9c-trichlorocamphen-2-one formed from toxaphene components in an anaerobic soil. J Agric Food Chem 45:4116–4121

3 Chiral Xenobiotics in the Environment

Neglect of stereoselectivity of biological objects and stereospecifity of bioactive agents nevertheless is a still persisting aspect of today's pharmacology and toxicology. It results in heavily biased "scientific" data such as half-life times, biological availabilities, concentration response relationships etc. for racemates and mixtures of isomers in general. It is like presenting the age or body weight of a married couple or even a four-person family (a mixture of 4 stereoisomers) in one figure. Such impressive nonsense is still rather common.

Ariëns, 1988 [150]

In the case of chiral pollutants, environmental studies have historically neglected to determine the adverse effects associated with particular enantiomers, and which enantiomers may persist in the environment. Consequently, much of the environmental data that has been collected for chiral pollutants, including many of the chemicals raising the greatest public concern, may have represented only the presence of relatively innocuous enantiomers.

Lewis et al., Oct. 1999 [164]

The authors of the present monograph would like to leave it to the reader's judgment as to whether the rapid development of the enantioselective analytical approaches and the results that have thus become accessible during the last decade, as summarised in the subsequent chapters, still justify the critical remarks cited above.

3.1
Microbiological Transformation of Chiral Environmental Xenobiotics

3.1.1
Laboratory Experiments

In 1992, Ludwig et al. [1] were the first to report the application of enantioselective capillary gas chromatography (cGC) to a microbial transformation experi-

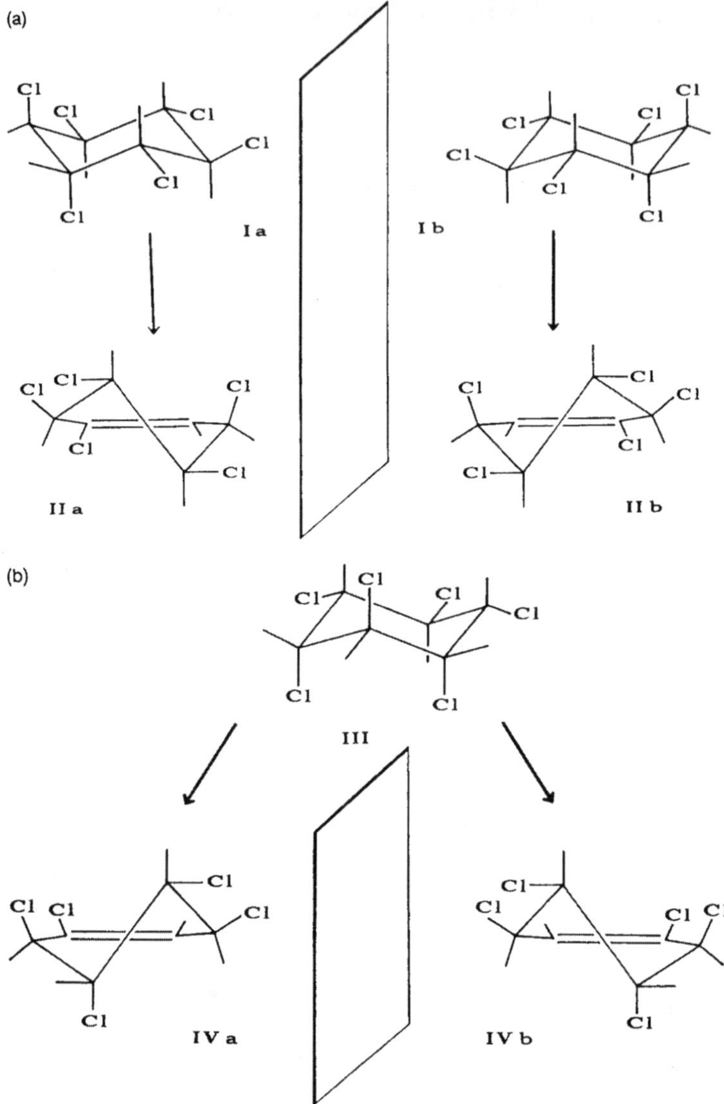

Fig. 3.1. *Top*: Enantiomers of α-HCH (Ia and Ib) and the corresponding β-PCCH enantiomers (IIa and IIb). *Bottom*: γ-HCH (III) and the corresponding γ-PCCH enantiomers (IVa and IVb); from [1]

ment that included the environmental pollutants α- and γ-hexachlorocyclohexane (α-HCH and γ-HCH, respectively, Figs. 1.5 and 3.1) as well as their transformation products β-pentachlorocyclohexene and γ-pentachlorocyclohexene (β-PCCH and γ-PCCH, respectively). α-HCH is the only chiral isomer of the eight conceivable HCH isomers (Fig. 3.1), while γ-HCH is achiral, but it is the only

HCH isomer that exhibits insecticidal activity. The transformation products of the two HCH parent compounds, β-PCCH and γ-PCCH, are chiral (Fig. 3.1).

In order to allow a comparison between the laboratory experiments and natural conditions in the open sea, the experiments were performed with a microbial community obtained from aerobic sediment and surface water off the North Sea island Heligoland (54° 11' N, 7° 53' E) and in the presence of additional carbon sources. Thus the formation of specific microbial cultures was avoided, and it was tentatively assumed that the original microbial community could be perpetuated at least for the experimental period. Furthermore, the cultures were kept under conditions which are known to promote the growth of marine bacteria much better than the growth of terrestrial bacteria, i.e., at room temperature and in salt water. In addition, microscopic examinations confirmed that both the size and the mobility of the microorganisms indicated the nearly exclusive presence of marine bacteria in the culture obtained. For the transformation experiment the sterile, aged seawater medium was inoculated with 0.2 mL of the enrichment culture, 2 mg of γ-HCH and 0.4 mg of α-HCH and shaken under the same conditions as the sterile control. For further experimental details the reader should refer to the original paper [1].

Basically, an enantioselective transformation of α-HCH can be investigated by determining the enantiomeric excess either of α-HCH or of its first transformation products throughout the experimental period. However, it turned out that during an experimental period of 4 weeks, the amount of α-HCH that had been transformed was less than 10%, i.e., the background level of technical (racemic) α-HCH with an enantiomeric ratio of 1:1 remained too high to obtain sufficiently reliable values to follow and verify an enantioselective process. Therefore, Ludwig et al. [1] suggested that the transformation process be investigated by continuously determining the enantiomeric excess of the transformation products of α-HCH. In the case of the prochiral γ-HCH, an enantioselective microbial transformation can be studied only by determination of the enantiomeric excess of its chiral transformation products, e.g., γ-PCCH.

As shown in Fig. 3.2a and 3.2b, the formation of β-PCCH by microbial transformation of α-HCH and of γ-PCCH by microbial transformation of γ-HCH, respectively, started about 4 days after inoculation. The diagrams represent the average values of two experiments. Both α-HCH and γ-HCH contained a small amount of PCCH, which caused a minimum background level in the sterile controls. The control experiments were performed at pH 7.5 and, in addition, at pH 8.5, because during the yeast extract digestion the pH value of the inoculated sample rose from 7.5 to 8.5 as a result of NH_3 production (for details see [1]). The experiment was stopped after 28 days, because the concentration of the PCCH isomers remained constant. This may have been caused by a reduced microbial activity and/or a simultaneously occurring PCCH degradation. This latter aspect will be discussed in detail below.

These results imply that the first step in the aerobic transformation of α-HCH by microorganisms is the energetically advantageous *trans*-HCl elimination, where each of the α-HCH enantiomers can be dehydrochlorinated to one specif-

SEPARATION OF ENANTIOMERS OF MARINE POLLUTANTS

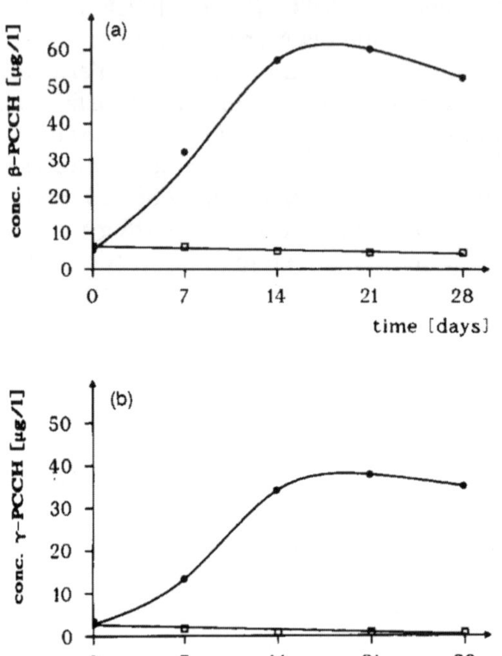

Fig. 3.2a, b. Formation of **a** β-PCCH during α-HCH transformation under aerobic conditions and **b** γ-PCCH during γ-HCH transformation under aerobic conditions. Average values of two experiments (*filled circles*) and of two sterile controls (*open squares*; pH 7.5 and 8.5); from [1]

ic β-PCCH enantiomer only (Fig. 3.1). The PCCH isomers were identified by comparison of the retention times on two capillary columns of different polarity and by cGC/MS.

Basically, the prochiral γ-HCH can form two γ-PCCH enantiomers, as shown in Fig. 3.1. During a nonenzymatic dehydrochlorination of the HCH isomers in an achiral environment, a racemate as reaction product is expected (see Sect. 3.4). However, in the case of an enzymatic elimination, one would predict a preferential formation of specific PCCH enantiomers. However, the results summarised in Table 3.1 reveal quite different answers for the enzymatic transformation of the chiral α-HCH and the prochiral γ-HCH parent compounds. Whereas the average enantiomeric ratios of β-PCCH changed from 1.18 at the beginning to a maximum value of 1.34 and then to 1.17 at the end of the experiment (enantioselectivity), they remained at about 1.00, within the limits of error, during the whole experiment in the case of γ-PCCH (no enantioselectivity).

In order to investigate the further fate of β-PCCH and to determine whether the maximum in the enantiomeric ratios which can be inferred from Table 3.1 is

Table 3.1. Enantiomeric ratios (defined as first-eluting peak divided by second-eluting peak areas) of β-PCCH and γ-PCCH formed by microbial transformation of α-HCH and γ-HCH, respectively, during a period of 4 weeks. The data are average values of two injections. The control data for β-PCCH showed values of 1.00±0.02 throughout the experiment; from [1]

Time (days)	Enantiomeric ratios			
	β-PCCH from α-HCH		γ-PCCH from γ-HCH	
	Experiment I	Experiment II	Experiment I	Experiment II
0	–	–	–	–
7	1.16	1.20	0.99	1.00
14	1.32	1.36	1.02	0.98
21	1.13	1.21	1.00	0.99
28	1.18	1.15	1.01	1.00

Table 3.2. Enantiomeric ratios (defined as first-eluting peak divided by second-eluting peak areas) of β-PCCH during a period of microbial transformation of 21 days, and of the sterile control (pH 7.5). The data are average values of two injections; from [1]

Time (days)	Enantiomeric ratios of β-PCCH		
	Experiment I	Experiment II	Control
0	1.00	1.00	1.00
3	0.99	0.97	1.00
6	0.85	0.83	1.00
9	0.77	0.79	0.99
12	0.71	0.76	1.00
16	0.63	0.71	0.99
21	0.50	0.51	1.00

caused by an enantioselective degradation of β-PCCH, Ludwig et al. repeated their microbial transformation experiment under the same experimental conditions, starting, however, from rac-β-PCCH [1]. On the basis of the results they obtained (Table 3.2), the appearance of a maximum of the enantiomeric ratios of β-PCCH during the microbial α-HCH transformation is conceivable. Obviously, the β-PCCH enantiomer that is being produced faster and in higher concentration is also degraded faster. Furthermore, it is evident that the decrease in the enantiomeric ratios of β-PCCH at the end of the α-HCH transformation (see Table 3.1) is caused by an increase in the more enantioselective and faster β-PCCH decomposition. An assignment of the peaks of the β-PCCH enantiomers was achieved by production of (–)-α-HCH according to the method of Cristol [3], and by its subsequent dehydrochlorination. Thus it was shown that (–)-α-HCH correlates with the second β-PCCH peak (according to the column used by Ludwig et al. [1]). This, in turn, implies that (+)-α-HCH and its corresponding β-PCCH enantiomer are degraded more easily. Therefore, Ludwig et al. tentatively assumed that the responsible enzymes prefer a common structural element represented by (+)-α-HCH and the corresponding β-PCCH enantiomer. It is worth noting that Vetter and co-workers meanwhile have been able to determine the direction of the optical rotation for the two β-PCCH enantiomers [4].

The laboratory transformation experiment performed by Ludwig et al. [1] with α-HCH and microorganisms isolated off the island of Heligoland reflect the microbial transformation which occurs in the open sea sufficiently well. This aspect will be pursued in the next section (Sect. 3.1.2). Furthermore, the results obtained by Ludwig et al. proved for the first time that the enantioselective gas chromatographic analysis of the enantiomeric ratios of chiral marine pollutants can be used as a tool to follow their microbial transformation pathways.

In addition to the nonpolar chlorinated hydrocarbons described above, several more polar organic pesticides play an important role in the marine ecosystem. As shown by Ludwig [5], phenoxyalkanoic acid herbicides may attain considerable concentrations in the German Bight, e.g., the achiral 2,4-dichlorophenoxyacetic acid (the so-called 2,4-D) in the order of 10 ng/L, the chiral 2-(2,4-dichlorophenoxy)propanoic acid (dichlorprop or DCPP, see Fig. 3.33) in the order of 2 ng/L, the chiral 2-(4-chloro-2-methylphenoxy)propanoic acid (MCPP) in the order of 11 ng/L, and another chiral isomer of MCPP in the order of 7 ng/L were found. In the case of the chiral phenoxyalkanoic acid derivatives, only the R-enantiomer is an active herbicide. In some commercially available mixtures, however, the racemate is prevailing. As a consequence, both the R- and the S-enantiomer are transported into the North Sea thus potentially influencing the marine ecosystem. Therefore, Ludwig et al. focused on the question as to whether or not a microbial marine culture is able to degrade the two enantiomers of chiral phenoxyalkanoic acid derivatives or, rather, is only one enantiomer accumulated in the marine ecosystem [6]. Polar organic compounds of this kind cannot be investigated directly by cGC; however, the application of high-performance liquid chromatography (HPLC) using α_1-AGP as chiral stationary phase (see Sect. 2.2.3) has turned out to be an elegant method both for the clean-up step and for the separation of the enantiomers of the chiral phenoxyalkanoic acid derivatives. For the microbial transformation experiment, Ludwig et al. used the same mixed culture of marine microorganisms as previously used for the α-HCH transformation experiments described above (for details see [6]).

Typical HPLC chromatograms obtained from a sterile control and a mixed microbial culture after a period of three weeks are shown in Fig. 3.3A and B, respectively, for DCPP. It is obvious that the R-enantiomer was degraded by the microbial culture, while the S-enantiomer appears to remain largely unaffected. The exclusive degradation product identified during the first period of the experiment was 2,4-dichlorophenol (2,4-DCP). This implies that as a first step the ether bond of the phenoxy group is broken thus giving rise to the formation of the respective halogenated phenol, similar to results reported for the degradation of 2,4-D by terrestrial microorganisms.

Quantitative insight into the temporal development of the concentrations of the DCPP enantiomers was gained during two experiments, each of three weeks

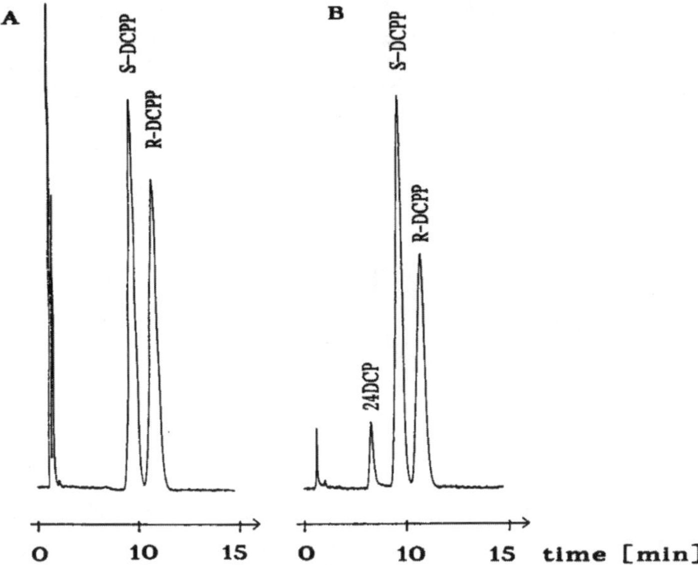

Fig. 3.3A, B. HPLC chromatograms of the enantiomers of chiral 2-(2,4-dichlorophe-noxy)propanoic acid (dichlorprop or DCPP) of **A** a sterile control and **B** after transformation by a mixed culture of marine microorganisms (21 days); α_1-AGP-column 100×4 mm; water/2-propanol 96:4 with 10 mM phosphate buffer; pH 4.85; velocity 0.9 mL/min; UV-detection at 240 nm; from [6]

duration. These results confirm the assumption stated above that the microbial marine culture which represents at least the south-eastern part of the North Sea (German Bight) is able to transform exclusively the R-enantiomer, while the concentration of the S-enantiomer remains constant within the error of this method. The authors assumed a statistical error of at maximum 2 to 3%, because the only preparation step prior to injection into the HPLC system comprised a dilution of the medium which has no effect on the ratio of the enantiomers. As a consequence of the exclusive transformation of the R-enantiomer, considerable shifts of the enantiomeric ratios S/R-enantiomer were observed during the experimental period of three weeks. The results, which are given for the two experiments and the sterile control in Table 3.3, illustrate the good reproducibility of this method. Whether the exclusive transformation of the R-enantiomer leads to an accumulation of the S-enantiomer in the marine ecosystem has to be investigated by additional *in situ* measurements. With regard to MCPP, the marine microbial culture used by Ludwig et al. was not able to degrade this compound. Further investigations will need to show whether this leads to accumulation effects of the latter phenoxyalkanoic acid derivative in the marine ecosystem.

The most notable difference between the results obtained for the microbial transformation experiments with *rac-α*-HCH and *rac*-DCPP is as follows: In the case of *rac-α*-HCH, *rac-β*-PCCH, and *rac-γ*-PCCH the mixed culture of marine

Table 3.3. Enantiomeric ratios (*S/R*) of 2-(2,4-dichlorophenoxy)propanoic acid (dichlorprop or DCPP) during a period of 21 days as encountered in the presence of a mixed culture of marine microorganisms and in a sterile control. The values, which were determined by HPLC using a chiral α_1-AGP-column, are averages of two injections; from [6]

Time (days)	Enantiomeric ratios (*S/R*)		
	Experiment I	Experiment II	Sterile control
0	0.99	1.00	0.99
3	1.02	1.01	0.99
7	1.08	1.05	1.01
10	1.14	1.11	0.99
14	1.34	1.35	1.00
17	1.37	1.39	1.00
21	1.41	1.41	1.00

microorganisms was able to transform the two enantiomers of these pollutants, although this was achieved at different rates. As a consequence, a shift of the enantiomeric ratios was observed. DCPP is also transformed enantioselectively; however, in contrast to the results with *rac*-α-HCH and the HCH metabolites, the marine microbial community exclusively transforms the *R*-enantiomer, while the *S*-enantiomer appears to be persistent.

Instead of applying enrichment cultures of marine microorganisms, Buser and Müller performed microbial transformation experiments with sludge from the anaerobic stabiliser of the communal sewage treatment plant at Zürich-Glatt, Switzerland [7]. This plant is considered to be typical of many other installations in Switzerland and Central Europe. They studied the transformation of the four most common hexachlorocyclohexane isomers, the chiral α-HCH, and the prochiral γ-HCH, δ-HCH, and β-HCH, in sewage sludge under anaerobic conditions. Approximately 250 g of sewage sludge in a 300 mL clear glass serum bottle was fortified with 100 µL of an ethyl ethanoate solution containing 400–500 µg of α-HCH, γ-HCH, or technical HCH. The bottles were tightly capped and incubated on a horizontal shaker at 298 K (25 °C) for up to 14 days in the dark. Samples were taken at different time intervals, extracted with *n*-hexane and analysed using enantioselective cGC/MS.

Significant transformation was observed for γ- and α-HCH with half-lives between 20.4 and 99 h. High enantioselectivity in the transformation of the α-HCH was indicated by significantly different rates for the (+)-enantiomer (20.2×10^{-3} h^{-1}) and the (–)-enantiomer (7.26×10^{-3} h^{-1}) resulting in an apparent enrichment of (–)-α-HCH in the digested samples [7]. Further details on the experimental results obtained by Buser and Müller can be found in Section 3.3.8.

It is noteworthy that Buser and Müller also tried to investigate the transformation products that were formed during their transformation experiment. In preceding model experiments they synthesised β-PCCH by dehydrohalogenation thus confirming the previous conclusions drawn by Ludwig et al. [1] and

Hühnerfuss et al. [2] with regard to the reaction mechanism. But though Buser and Müller could thus reliably verify their experimental approach, they were unable to detect β-PCCH in any of the sewage sludge samples spiked with α-HCH. Therefore, they conjectured that the metabolite initially formed is further degraded at rates that are significantly faster than the rate of their formation. This hypothesis is largely in line with the observations made by Ludwig et al. [1] with regard to the formation and degradation rates inferred from transformation experiments with marine microorganisms under aerobic conditions, although, under the latter experimental conditions, β-PCCH could be analysed.

An experimental approach that moved somewhere near the borderline between microbial and enzymatic transformation of environmental pollutants was reported by Garrison et al. [68]. They investigated the chiral compound o,p'-DDT, which comprises 12 to 20% of technical grade DDT. This isomer is presumed to be a human endocrine disruptor because of its strong estrogenic activity in rats. It is found in the environment wherever p,p'-DDT was applied or is being found due to atmospheric transport processes. Both DDT isomers are microbially transformed by similar pathways: anaerobically to o,p'- and p,p'-DDD and o,p'- and p,p'-DDE or aerobically to o,p'- and p,p'-DDA [2-(2-chlorophenyl)-2-(4-chlorophenyl)ethanoic acid and 2,2-bis(4-chlorophenyl)ethanoic acid, respectively]. Microbial transformation of these compounds appears to be very slow and is expected to be enantioselective. However, reports dealing with the chirality of o,p'-DDT in the environment are scarce. These include observations of the occurrence of unequal concentrations of the enantiomers of o,p'-DDT in soil extracts (see Sect. 3.5.2; [69]) and in cod liver oil (see Sect. 3.2.1.1; [26]).

Garrison et al. were mainly interested in o,p'- and p,p'-DDT degradation by plant enzymes as a potential phytoremediation process. Their experimental approach is based on the observation that several plant enzymes can degrade organic pollutants [68]. These include, for example, nitroreductases that reduce nitro moieties to amines and dehalogenases that replace halogen with hydrogen. For instance, aliphatic halocarbons such as hexachloroethane and carbon tetrachloride are readily reduced to compounds with a lower degree of halogenation by dehalogenase enzymes that occur in a variety of plants. Some of these enzymes have been extracted and partially characterised. However, the authors cannot fully exclude a priori that a concurrent microbial enzymatic degradation may take place.

The plant degradation experiments were carried out as follows: 20 g of Elodea (*Elodea canadensis* L.), collected from a local lake, were rinsed with distilled water and placed in each of a series of serum bottles containing 100 mL of the boiled aqueous extracts of a sample of the lake sediment. The water in the bottles was dosed with 1 µg/mL each of o,p'- and p,p'-DDT, and the bottles were sealed. The combination of *Elodea* and water in the sample was blended and extracted with a 1:1 mixture of n-hexane and acetonitrile, evaporated to dryness, taken up in methanol, diluted with water and concentrated on a C-18 solid phase extraction tube. The methanol eluate was analysed for the DDT congeners by cGC using modified cyclodextrins as chiral selector and for acidic degradation products, such as DDA, by enantioselective capillary zone electrophoresis (CZE). Garrison et al. used a permethyl-trifluoroacetoxypropyl-γ-cyclodextrin phase which provided baseline separation of the enantiomers of o,p'-DDT and, for the needs of their experiment, adequate separation of the enantiomers of o,p'-DDD.

The reductive dehalogenation products, o,p'- and p,p'-DDD, began to appear as soon as the parent compounds started to decay, where the p,p'-DDT degraded slightly faster than the o,p'-isomer. Both DDD products slowly decayed from the reaction solution over several weeks. Plots of the loss of o,p'-DDT and p,p'-DDT matched that for first-order kinetics; both compounds were completely lost from the *Elodea*/water reaction medium in from 5 to 24 days, depending apparently on the growth conditions of the plants when harvested for use in the six kinetic runs.

It turned out that the transformation of o,p'-DDT and formation of o,p'-DDD were *not* enantioselective. For both compounds, each of the two enantiomers was always of the same concentration, at least within experimental error, throughout the course of the reduction reaction. Garrison et al. speculated that the distance of the reactive centre from the chiral centre might preclude enantioselectivity, but we are not in favour of this explanation, because the chiral centre is assumed to be sufficiently close to induce an enantioselective preference. Thus, further investigations are needed to elucidate the enantioselective enzymatic degradation of o,p'-DDT and its metabolites.

A partial microbial reduction cannot *a priori* be ruled out, although it would presumably lead to an enantiomeric excess. In order to clarify this point, Garrison et al. carried out a cobalt-60 gamma irradiation test in parallel experiments. The irradiated samples were reduced faster than the nonirradiated controls, whereas they should have reacted much slower if microbial degradation were occurring in the nonirradiated samples. Therefore, the authors conclude that microbial degradation in this particular case plays an immeasurable role.

3.1.2
In Situ Investigations in Marine and Limnic Waters

In 1991, Faller et al. [8] raised the question "Do marine bacteria degrade α-HCH stereoselectively?", and they were the first to answer this question in the affirmative by applying cGC with heptakis(3-*O*-butyryl-2,6-di-*O*-*n*-pentyl-β-cyclodextrin as a chiral stationary phase to residual analysis of environmental samples. In a subsequent systematic investigation by Faller et al. [9], 16 water samples representing all parts of the North Sea that are interesting both from an oceanographic and chemical view were taken by means of 10 L glass samplers. The sampler consists of an all-glass round flask (flat 100 mm wide ground neck) that is fixed by a stainless-steel protective basket and closed by a stainless-steel cover with two inlets for glass tubes which can be opened at a predetermined water depth. For further details the reader should refer to the paper by Gaul and Ziebarth [10]. In general, the sampling depth was 10 m. The water samples were extracted, purified, and analysed according to known procedures, which include

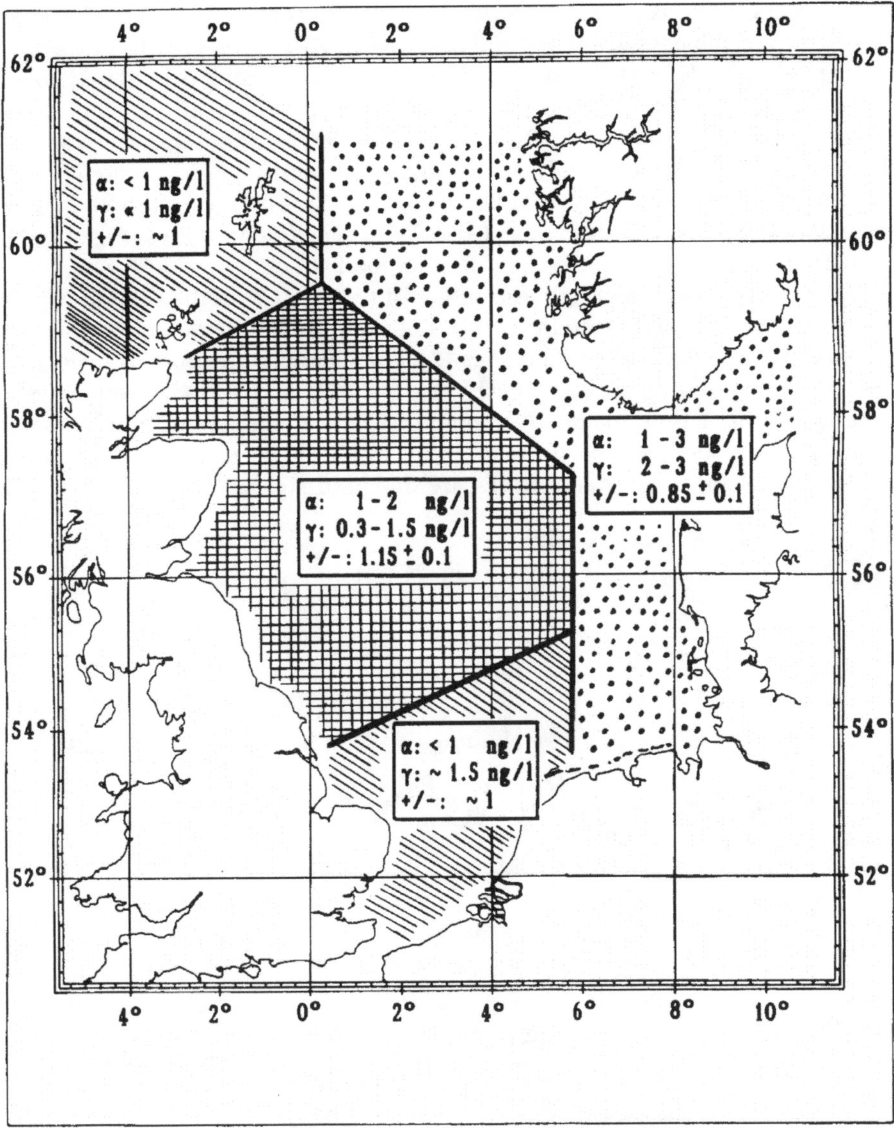

Fig. 3.4. Diagrammatic representation of the three regions in the North Sea exhibiting different relations of (+)-α-HCH/(−)-α-HCH (~1; <1; >1). The total number of stations was not large enough to calculate exact zones by means of interpolation programs; from [9]

extraction of 10 L of the respective seawater sample by 200 mL n-hexane, purification of the extract by column chromatography over an Al_2O_3 column, and fractionation by HPLC (further details in [2]). Subsequently, the thus purified fraction was analysed with regard to enantiomeric ratios (ER) of the α-HCH enantiomers applying enantioselective cGC. The ER is directly obtained by peak integration of the gas chromatogram and dividing the peak area E_1 of the first eluting $(+)$-α-HCH by the peak area E_2 of the second eluting $(-)$-α-HCH enantiomer, i.e., $ER = E_1/E_2$. The results thus obtained revealed different microbial transformation pathways in the North Sea (Fig. 3.4): while in the eastern part of the North Sea, including the German Bight and the Skagerrak, a preferential transformation of the $(+)$-α-HCH was observed (ER ~0.85), in the area east off the coast of Great Britain preferably $(-)$-α-HCH is degraded (ER ~1.15) and $(+)$-α-HCH appears to be more resistant to microbial attack.

In 1992, the preferential transformation of $(+)$-α-HCH in the eastern part of the North Sea was confirmed by analyses of additional water samples from the German Bight and from the Baltic Sea [2]. Furthermore, the main transformation products both of α- and γ-HCH, i.e., β-PCCH and γ-PCCH, respectively, were for the first time included in enantioselective analyses of marine water samples. The respective gas chromatograms of these two HCH metabolites, extracted from Baltic sea water samples, are shown in Fig. 3.5.

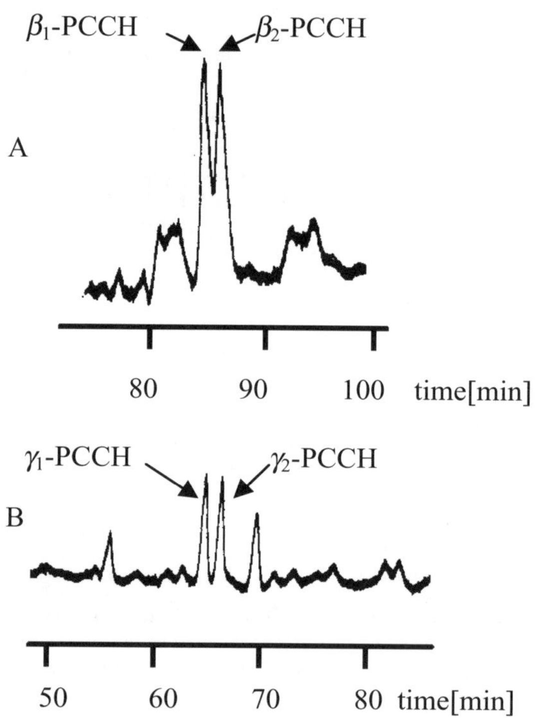

Fig. 3.5 A, B. Chromatograms of **A** the two β-PCCH and **B** the two γ-PCCH enantiomers extracted from Baltic Sea water samples (A, stations 1–10 pooled; B station 4, see Fig. 3.6) using a fused-silica capillary column coated with 50% heptakis(2,3,6-tri-O-n-pentyl)-β-cyclodextrin and 50% OV-1701. Column temperature program: 323 K increased at 10 K/min to 388 K; carrier gas 45 kPa helium; on-column injection; ECD; after [2]

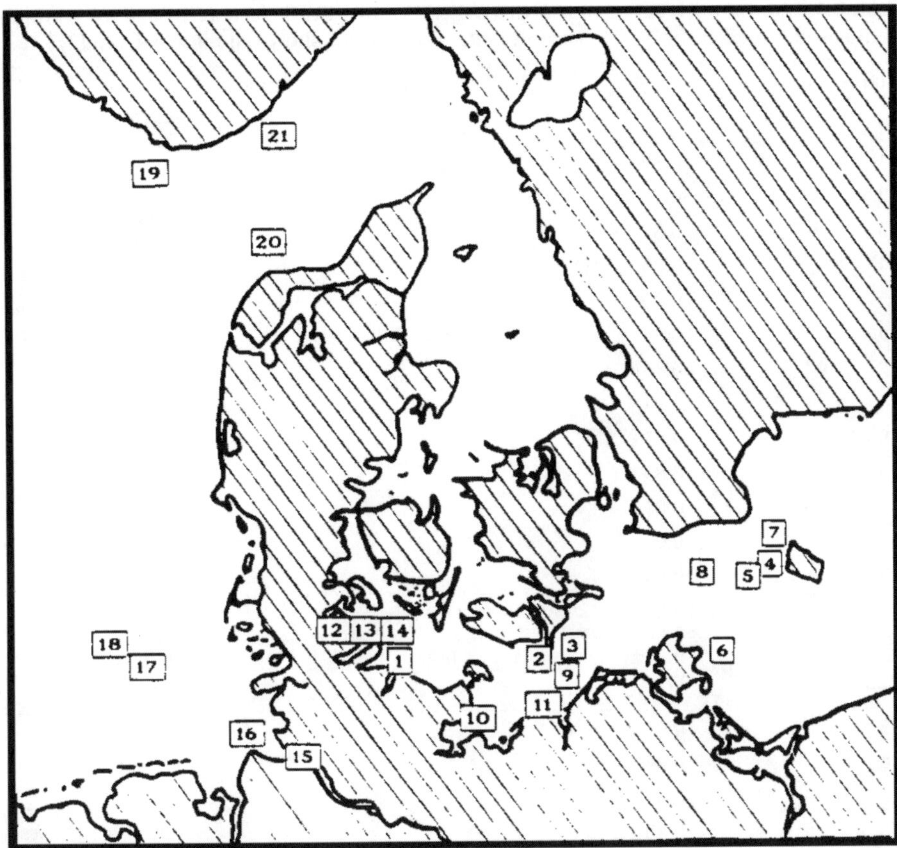

Fig. 3.6. Positions of the 21 stations at which water samples were taken for the enantioselective analyses of α-HCH, β-PCCH and γ-PCCH; from [2]

For α-HCH, the average ER value calculated from the water samples, which were taken at the stations depicted in Fig. 3.6, turned out to be 0.85±0.03 for the Baltic Sea and 0.87±0.05 for the North Sea (Table 3.4). Furthermore, a verification of the conclusions drawn from the North Sea results was accomplished by comparison with the systematic laboratory measurements described in Section 3.1.1.

Meanwhile, Bidleman and co-workers carried out additional investigations on the enantioselective microbial transformation of α-HCH in lakes and coastal bays [11–14], and in marine waters of other regions of the world's oceans [15, 16]. These studies formed the basis for gaining deeper insight into air/water exchange processes, which will be further discussed in Section 3.6. In this chapter, the experimental evidence for microbial transformation of pollutants in lakes and marine waters as inferred from enantioselective cGC results will be summarised.

Table 3.4. Enantiomeric ratios of α-HCH, β-PCCH and γ-PCCH as determined for seawater samples obtained at the North Sea and Baltic Sea stations shown in Fig. 3.6; from [2]

Station No.	Enantiomeric ratios		
	$(+)$-$/(-)$-α-HCH	β_1-$/\beta_2$-PCCH	γ_1-$/\gamma_2$-PCCH
1	0.89	a	1.16
2	0.79	a	1.16
3	0.87	a	1.17
4	0.85	a	1.12
5	0.83	a	
6	0.85	a	
7	0.83	a	b
8	0.84	a	b
9	0.85	a	1.13[b]
10	0.83	a	b
11	0.85		
12	0.92		
13	0.84		
14	0.82		
Average value Baltic Sea	0.85±0.03	0.97[a]	1.15±0.02
15	0.81		
16	0.84		
17	0.87		
18	0.83		
19	0.88		
20	0.94		
21	0.94		
Average value North Sea	0.87±0.05		

[a] Value of pooled stations 1–10.
[b] Value of pooled stations 7–10.

The enantiomeric ratios of the α-HCH enantiomers determined by Bidleman and co-workers in lake waters were largely comparable with those reported by Hühnerfuss and co-workers for the eastern part of the North Sea and the Baltic Sea, i.e., a preferential microbial transformation of the $(+)$-α-HCH was observed in all cases. The ER values determined in the Arctic "Amituk Lake", N.W.T. in Canada, ranged between 0.65 and 0.99, while for Lake Ontario an average value of 0.85 and for Resolute Bay of 0.93 were obtained.

A more complex situation was encountered during a cruise in northern marine waters [16]. In the summers of 1993 and 1994, seawater samples from the

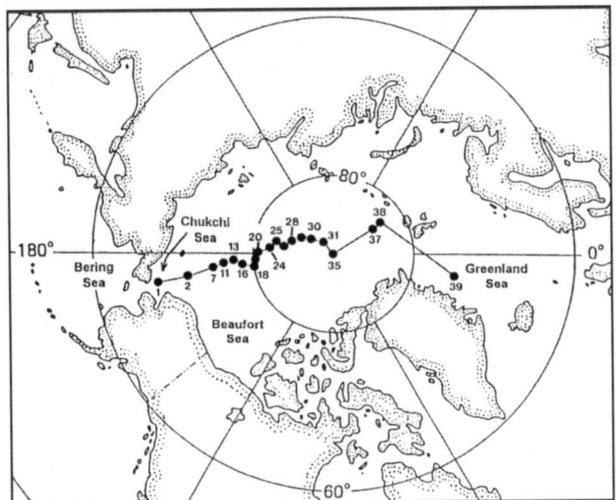

Fig. 3.7. Cruise track of AOS-94. Dots running from the Chukchi Sea to the Greenland Sea correspond to the station numbers in Table 3.5; from [16]

surface layer (40–60 m) were collected to determine the spatial distribution of α-HCH and organochlorine pesticides on expeditions that crossed the Arctic Ocean from the Bering and Chukchi Sea to the North Pole, to a station north of Spitsbergen, and then south into the Greenland Sea (Fig. 3.7). The enantiomeric ratios of dissolved α-HCH in arctic waters were generally >1.00 in the Bering-Chukchi Seas, while depletion of the (+)-α-HCH was found in the Arctic Ocean and Greenland Sea, with ERs <1.00 (Table 3.5). Similar to the results reported by Faller et al. for the western and eastern parts of the North Sea, different microbial transformation pathways can be assumed for the Bering-Chukchi Seas and for the Arctic Ocean: in the former case preferential microbial degradation of (–)-α-HCH was observed which implies that (+)-α-HCH appears to be more resistant to microbial attack (ER >1), while in the latter case a preferential transformation of the (+)-α-HCH was determined (ER <1.00).

Furthermore, Jantunen and Bidleman also analysed the marine water samples collected during the AOS-94 cruise for the enantiomeric ratios [(+)/(–)-enantiomers] of *trans*-chlordane, *cis*-chlordane and heptachlor *exo*-epoxide, a major metabolite of heptachlor (Table 3.5). The ER values determined for heptachlor *exo*-epoxide ranged between 1.47 and 1.76, which implies a preferential formation of the (+)-enantiomer. The range of ERs was tight considering the wide expanse covered by the stations. The two chlordane isomers, *trans*- and *cis*-chlordane, were close to racemic in the dissolved phase (ER=0.94–1.06).

Although HCHs were >99% in the dissolved phase at most stations, levels of HCHs in water were high enough to allow ER values to be measured on the filters of the large volume samples. The particulate α-HCH showed the same, or greater, enantioselective transformation than the dissolved fraction (Table 3.5). Both

Table 3.5. Enantiomeric ratios [(+)-/(–)-enantiomer] of α-HCH, heptachlor *exo*-epoxide (HEPX), *trans*-chlordane (TC), and *cis*-chlordane (CC) in surface water samples collected at the stations shown in Fig. 3.7; two different chiral phases were applied: BGB-172 and Beta-DEX, respectively; from [16]

Station No.	α-HCH/BGB-172		α-HCH/Beta-DEX		HEPX	TC	CC
	Dissolved	Particle	Dissolved	Particle	Dissolved	Dissolved	Dissolved
Standard	1.00		0.99		1.01	0.99	0.99
1	1.11	1.27	1.08	1.26	1.52	1.00	1.00
2	1.09	1.05	1.09	1.03	1.47	1.00	1.04
11	1.06	na	1.05	0.98	1.64	1.00	1.06
13	0.95	0.98	0.89	0.93	1.58	1.06	1.02
16	na	1.10	0.83	1.07	na	0.98	na
20	na	0.21	0.82	0.18	na	0.99	na
24	0.92	0.88	0.87	0.95	1.76	0.97	0.99
25	0.96	0.93	0.89	0.95	1.59	0.98	1.01
28	0.92	0.91	0.87	0.89	1.59	1.01	0.94
29	0.90	1.16	0.88	1.23	1.65	0.97	na
31	0.96	0.42	0.90	0.42	1.67	1.01	na
35	0.91	0.84	0.88	0.86	1.63	0.99	1.01
37	0.85	na	0.78	na	1.56	0.97	0.98
39	0.71	0.80	0.68	0.79	1.67	1.03	1.01

na=not analysed

fractions were depleted in the same enantiomer at most stations, but there appeared to be no relationship between the two, and reversal in the depleted enantiomer was found at stations 16 and 29. Some particulate samples showed enhanced transformation of α-HCH compared with the dissolved phase. For example, the ER of dissolved α-HCH at station 20 was 0.82, whereas the ER on the filter was 0.18. The ERs for dissolved α-HCH generally decreased with depth. At stations 37 and 38 north of Spitsbergen, surface ERs were 0.82, decreased to 0.64 at 109 m, 0.45 at 235 m, and 0.14 at 762 m. Jantunen and Bidleman offer two explanations for the greater enantioselectivity with depth: As particles settle from the surface, the sorbed α-HCH is metabolised and released back into the dissolved phase. Alternatively, the ERs may be typical of older, Atlantic-layer water, which lies below the pycnoline.

Recently, Franke et al. showed that the application of enantioselective gas chromatography to water sample extracts is not at all confined to the determination of classical organohalogen xenobiotics. They investigated the chlorinated bis(propyl) ethers (Cl_xBPE; x=2–4), an important new class of environmental contaminants in the River Elbe [60]. Quantitative analysis during the period 1992 to 1995 revealed total concentrations of up to 30 µg/L close to the Czech border and a gradual decrease to 4–2 µg/L towards the central part of the river. The

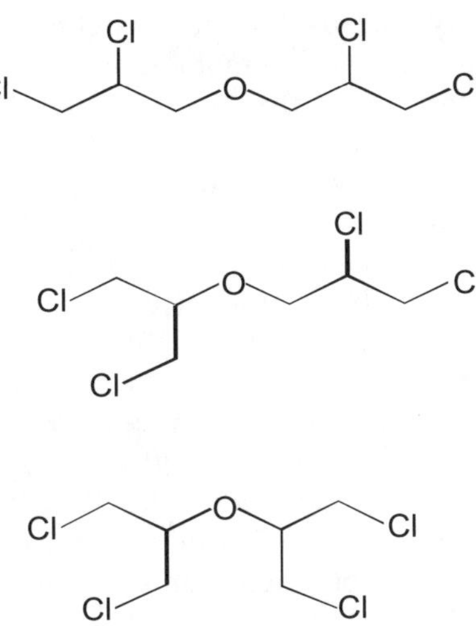

Fig. 3.8. Structures of three Cl$_4$BPEs. *1* Bis(2,3-dichloro-1-propyl) ether (2,3,2',3'-Cl$_4$BPE), *2* 1,3-dichloro-2-propyl-2,3-dichloro-1-propyl ether (1,3,2',3'-Cl$_4$BPE), and *3* bis(1,3-dichloro-2-propyl) ether (1,3,1',3'-Cl$_4$BPE)

authors conjectured that this change in concentration cannot solely be attributed to the diluting effect caused by uncontaminated tributaries, because their water supply is insufficient to account for this. The source of Cl$_4$BPE (Fig. 3.8) was a production site for epichlorohydrin close to the Czech border emitting considerable amounts of these compounds as side products.

During the investigation period, three Cl$_4$BPE isomers were present in an almost constant pattern over the upper and central part of the river, which corresponded to the pattern of the emitted derivatives (Fig. 3.8): 1,3,1',3'-Cl$_4$BPE (the digits indicate the position of the chlorine atoms; this simplified nomenclature is unambiguous) approx. 10–15%, the 2,3,2',3'-Cl$_4$-isomer approx. 28–35%, and 1,3,2',3'-Cl$_4$BPE approx. 55–57%. A shift in the isomeric pattern was, however, verified for the lower reaches of the river: 1,3,1',3'-Cl$_4$BPE increased in concentration, while 2,3,2',3'-Cl$_4$BPE decreased. Basically, the shift in the isomeric proportions towards the lower reaches of the river may result from a discrimination of the 2,3,2',3'-Cl$_4$BPE by physical or biochemical processes. A decision about this alternative was expected by an investigation of the enantiomeric composition over the entire length of the river, since 2,3,2',3'-Cl$_4$BPE is a chiral compound. A change in the enantiomeric ratios would be indicative of enzymatic microbial degradation, while retention of a racemic composition would point to a nonenzymatic process.

Appropriate chiral selectors for solving such problems often are modified cyclodextrin phases. However, the separation potential of a cyclodextrin-type chiral stationary phase is difficult to predict and, therefore, needs to be tested empirically. In this case, several cyclodextrin derivatives had to be evaluated with respect to their selectivity towards the chiral bis(chloropropyl) ethers with mixtures of standard compounds of different degrees of chlorination. In this connection, it should be emphasised that we have often had to carry out similar time-consuming tests when our attention has been drawn to new chiral environmental pollutants, e.g., bromocyclen or the polycyclic musk fragrances, which will be discussed in Sections 3.2.1 and 3.3. In order to illustrate the situation that might be the result of such endeavours, the separation potential of some cyclodextrin phases with regard to the chloro-bis(propyl) ether is as follows [60]: a satisfactory enantiomeric separation of 1,2'-Cl$_2$BPE and 1,3,2'-Cl$_3$BPE is achieved with octakis(3-O-butyryl-2,6-di-O-n-pentyl)-γ-cyclodextrin (Lipodex E). The stereoisomers of the tetrachloro compounds are not resolved on this phase. Heptakis(6-O-$tert$butyldimethylsilyl-2,3-di-O-methyl)-β-cyclodextrin resolves 1,3,2'-Cl$_3$BPE and the tetrachloro compounds 1,3,2',3'-Cl$_4$BPE and 2,3,2',3'-Cl$_4$BPE with the (S,S)-enantiomer of 2,3,2',3'-Cl$_4$BPE and the (+)-enantiomer of 1,3,2',3'-Cl$_4$BPE coeluting. Similarly, separations can be achieved with a heptakis(6-O-$tert$-butyldimethylsilyl-2-O-methyl-3-O-n-pentyl)-β-cyclodextrin column as long as a slower temperature program is applied. In addition, 1,2',3'-Cl$_3$BPE is completely separated into four stereoisomers. No separations of haloethers were observed on the column with the mixed phase of heptakis(2,6-di-O-methyl-3-O-n-pentyl)-β-cyclodextrin and heptakis(6-O-methyl-2,3-di-O-n-pentyl)-β-cyclodextrin. None of the investigated cyclodextrin phases was able to resolve all compounds of interest. However, since Franke et al. already knew from earlier investigations that the contamination of the River Elbe by tetrachloro compounds was more severe than with tri- and dichloro homologues, they focused on the investigation of the tetrachloro bis(propyl) ethers, and, as a consequence, they chose the heptakis(6-O-$tert$-butyldimethylsilyl-2,3-di-O-methyl)-β-cyclodextrin column which separated these homologues well. The assignment of the order of elution of the enantiomers of 2,3,2',3'-Cl$_4$BPE was achieved by enantioselective synthesis of the (2R,2'R/S)-stereoisomer. The determination of the absolute configuration of one of the two stereogenic centres is sufficient, since the *meso*-form (note: 2R,2'S=2S,2'R) can be easily detected in the gas chromatogram of the racemate by its peak height which is approximately twice that of the enantiomers of the chiral stereoisomer. The enantiomers of 1,3,2',3'-Cl$_4$BPE could be correlated by coinjection with an enantiomerically enriched sample (40% ee), which was obtained by preparative gas chromatography of the racemic compound.

Although coelution of the (S,S)-isomer of 2,3,2',3'-Cl$_4$BPE and the (+)-enantiomer of 1,3,2',3'-Cl$_4$BPE was encountered, a quantitative determination of the enantiomeric proportion was possible by virtue of the specific mass spectrometric fragmentation of the isomers, using specific ion traces. For a detailed discussion on this aspect, the reader should refer to [60]. The results thus obtained are summarised in Table 3.6. The enantiomeric ratios of 1,3,2',3'-Cl$_4$BPE showed values between 0.72 and 1.20. In the samples P1, P5, and P7 (see map in Fig. 3.9) a slight shift in the ER towards the (+)-isomer was observed. This trend was even more obvious in the samples from September 1995. A shift of the ER of the

Table 3.6. Enantiomeric ratios of chiral tetrachloro bis(propyl) ethers in the River Elbe at the stations shown in Fig. 3.9; from [60]

Sample No.	Location	1,3,2',3'-Cl$_4$BPE (−)/(+)	2,3,2',3'-Cl$_4$BPE (R/R')/(S,S')	meso/(S/S')
		February 1995		
	Bilina	1.02	0.94	1.91
P1	Schmilka	0.85	0.63	1.66
P3	Magdeburg	0.92	1.05	2.06
P5	Schnackenburg	0.89	0.78	1.87
P6	Zollenspieker	1.01	0.94	1.92
P7	Seemannshöft	0.87	0.41	1.16
P8	Grauerort	1.05	1.02	2.01
P9	Cuxhaven	1.20	1.59	2.88
		September 1995		
P1	Schmilka	0.86	0.54	0.74
P2	Zehren	1.00	0.83	0.83
P3	Magdeburg	0.84	0.53	1.75
P4	Tangermünde	0.82	0.44	1.42
P5	Schnackenburg	0.72	0.27	0.92

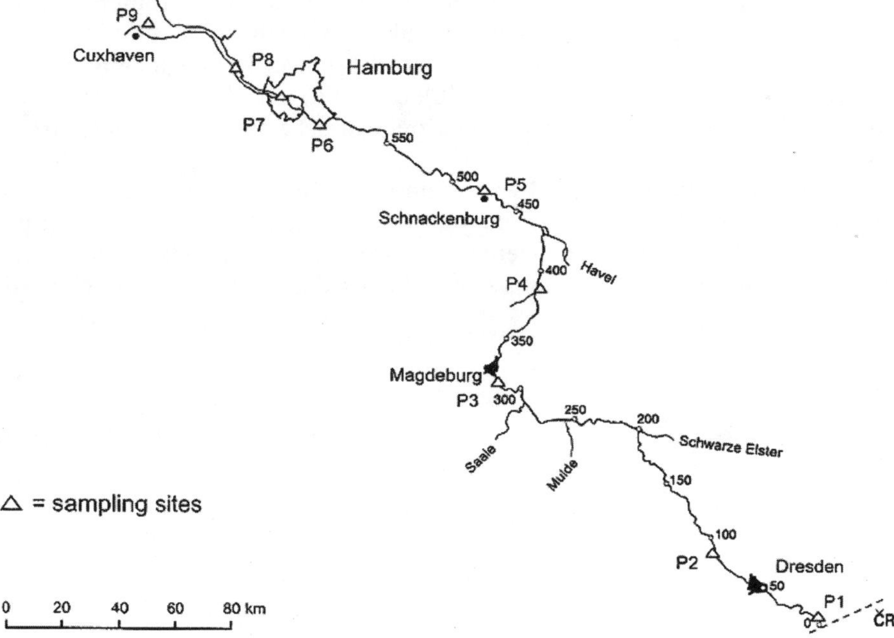

Fig. 3.9. Sampling sites P1 to P9 along the River Elbe during the campaign carried out by Franke et al. [60]

(–)-enantiomer to 0.72 (P5) was far beyond the medium deviation of the experimental method. Significantly different were the proportions in sample P9 with an increase in the ER of the (–)-enantiomer to 1.20.

The enantiomeric ratios of 2,3,2',3'-Cl$_4$BPE show a significant discrimination of the (*R,R*)-enantiomer in the samples "Schmilka", "Schnackenburg", and "Seemannshöft" (February 1995), as well as in the samples from September 1995 with ER values as low as 0.27. This shift of the enantiomeric ratios in the middle part of the River Elbe must be attributed to a decrease in the concentration of (*R,R*)-2,3,2',3'-Cl$_4$BPE. Due to the relatively low levels of these compounds in the series taken in September 1995, the enantioselective discrimination and the concomitant shift in the enantiomeric ratios can be clearly observed, since it is not superimposed by continuous inflow of racemic material. In addition, the biological activity may have been larger during the summer months, which may also have given rise to a more intense enantioselective discrimination. Similar to the values for 1,3,2',3'-Cl$_4$BPE, the ER values for the samples "Magdeburg", "Zollenspieker", and "Grauerort" were in the range of the racemate, considering the variability of the method. Again, the sample "Cuxhaven" with an ER value far above 1 deviates significantly from the other samples. Its chloroether concentration was relatively high, but Franke et al. could not offer a straightforward explanation for this observation.

In recent years, a new environmental contaminant, the insecticide bromocyclen, has been increasingly found in biota [49, 50, 66]. This brominated and chlorinated bicycloheptene contact insecticide (tradename Bromodan, Alugan®) is the Diels-Alder adduct of hexachlorocyclopentadiene and allyl bromide. It possesses neither a plane nor a centre of symmetry and, therefore, it exists in two enantiomeric forms (Fig. 3.10). Due to its very low mammalian toxicity, bromocyclen is currently used in Europe against ectoparasites for the treatment of domestic animals.

Bethan et al. [49] collected 10 water samples from the River Stör, a tributary of the River Elbe in northern Germany, and one water sample from the River Elbe close to the mouth of the River Stör. Moreover, samples from the influent as well as from the effluent of various sewage plants along the river were collected.

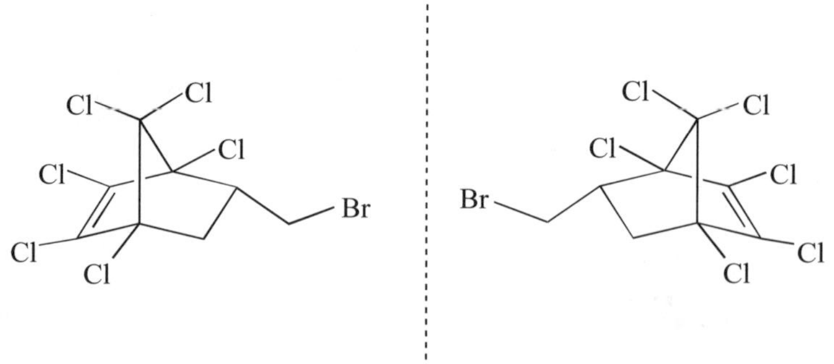

Fig. 3.10. Enantiomers of bromocyclen

Particular emphasis was placed on the question as to how bromocyclen gets into the River Stör and if the enantiomeric ratios determined in the fish samples correlate with the concentrations. The latter aspect will be further discussed in Section 3.2.1.1.

Close to the spring of the River Stör, concentrations of 37 pg/L were measured, followed by strong increases to 213 pg/L (near Neumünster) and 261 pg/L (near Itzehoe) presumably due to effluents of communal wastewater treatments plants, and a decrease to 51 pg/L in the course of its downstream movements towards the estuary, obviously caused by dilution effects. As the "hot spots" in the River Stör pointed to two local wastewater treatment plants as potential sources for bromocyclen, samples from their influents as well as from their effluents were analysed. The concentrations in 1995, which varied between 3.3 and 11.5 ng/L, were up to 100-fold higher than the increases in concentration that were found in the river. For example, an increase of 104 pg/L in the bromocyclen contamination of the Stör water, from the station upstream of the sewage plant of Itzehoe to the next station downstream of the sewage plant, reflects a concentration of 11.5 ng/L in the effluent of the sewage treatment plant.

With regard to the analysis of the enantiomeric excesses of bromocyclen in the Stör and Elbe water samples, no significant differences in the concentrations of the two enantiomers were found [49]. The enantiomeric ratios of all nine water sample extracts ranged between 1.01 and 1.05. This nearly racemic distribution indicates that bromocyclen is not or only insignificantly transformed by microbial processes in the river water.

Another new class of chiral environmental pollutants, the polycyclic musk compounds, was recently successfully investigated with regard to enantioselective transformation processes. A very comprehensive study in a wastewater treatment plant in Northern Germany included the compartments water, sediment, mussels, semipermeable membrane devices (SPMD), and different fish species from the pond of the plant. Because of the close correlation between these compartments, the results will be discussed in a separate section (see Sect. 3.3).

3.2
Transformation/Accumulation of Chiral Xenobiotics in Biota

3.2.1
Enzymatic Transformation Processes

3.2.1.1
Marine and Limnic Ecosystem

In 1991, Kallenborn et al. were the first to report the successful application of enantioselective cGC to biota extracts with the aim of investigating the enantioselective metabolism of α-HCH in organisms of different trophic levels [17]. The authors obtained common Eider ducks (*Somateria mollissima* L.) from the Oehe/Schleimünde wildlife refuge on the Baltic coast of Germany. Organ samples were taken only from healthy animals that, upon diving, had become trapped in the fishing nets of local fishermen and were thereby drowned. The Eider duck was chosen, because it largely favours blue mussels (*Mytilus edulis* L.) in its diet [163]. Blue mussels, in turn, are capable of strongly enriching pollutants and thus serve as "indicator organisms" to provide insight into the state of an aquatic environment. Thus, one of only a few examples is encountered where a simple "food chain" can be assumed (water→mussel→Eider duck). Normally, one should rather consider a "food web".

The tissue samples (liver, kidney, muscle) were homogenised with a threefold amount of anhydrous sodium sulphate followed by addition of the standard ε-HCH and extraction in a Soxhlet apparatus with *n*-hexane. The detailed procedure including the clean-up and fractionation steps can be found in the original paper [17]. Figure 3.11 shows the gas chromatograms of the α-HCH enantiomers for the complete food chain, i.e., water→blue mussel→common Eider duck (liver), using a modified cyclodextrin phase as chiral selector [18]. Whereas the blue mussel largely reflects the characteristics of the adjacent water area, i.e., an excess of the (–)-α-HCH, in the liver of the common Eider duck the (+)-α-HCH is dominant.

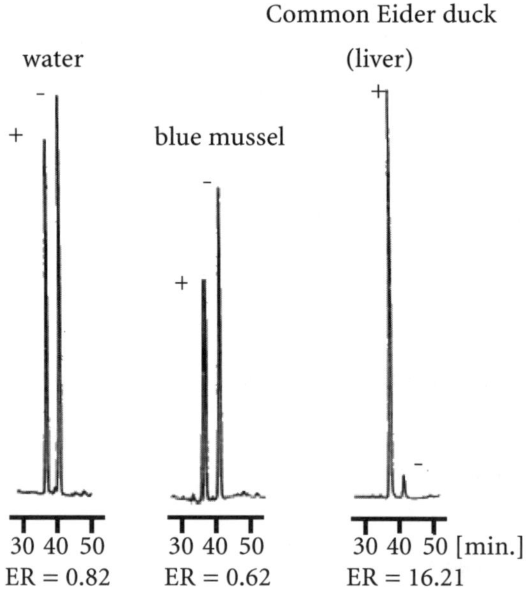

Fig. 3.11. Enantiomer separation of α-HCH extracted from a Baltic Sea water sample, from a mussel (*Mytilus edulis* L.) and from the liver of a common Eider duck (*Somateria mollissima* L.) using a fused-silica capillary column coated with 50% heptakis(2,3,6-tri-*O*-*n*-pentyl)-β-cyclodextrin and 50% OV-1701. Column temperature programme: initial 323 K, increased at 10 K/min to 388 K; carrier gas, helium (45 kPa); on-column injection; ER= enantiomeric ratio [(+)-/(–)-enantiomer]; from [18]

Detailed analyses of the extracts of all three common Eider duck tissues revealed that (+)-α-HCH was clearly enriched; almost enantiomerically pure (+)-α-HCH was present in the liver extracts. The enantiomeric purity of (+)-α-HCH isolated from liver extracts was so high that, after purification by HPLC, it could be used directly in model experiments. By contrast, the enantiomeric ratio (+)/(−)-α-HCH was about 7 in muscle extracts and about 1.6 in kidney extracts, whereby the values for these organs were slightly larger or smaller for different common Eider ducks. In this first study, the organs from a total of six common Eider ducks were investigated, so that the results can be considered sufficiently reliable.

An exact explanation for the appearance of different enantiomeric ratios of (+)-α-HCH in the organs of common Eider ducks cannot be given at the present time. It may be assumed, however, that the reason lies in the different physiological functions of the organs. For muscle and kidney, whose main functions are "locomotion" and "excretion", respectively, the content of extractable lipids is about 2%; these organs can, therefore, store lipophilic pollutants. Liver, which contains about 2.5% of extractable matrix, serves as a "detoxification organ" and, therefore, is not only capable of storing toxic compounds, but can also metabolise them to substances that the body can tolerate or excrete. Since the (+)-α-HCH found in the liver of common Eider ducks is almost enantiomerically pure, (−)-α-HCH is presumably more readily transformed enzymatically than the (+)-enantiomer in the liver. Although such a nearly enantioselective transformation had already been observed previously for biogenic organic compounds, the study by Kallenborn et al. [17] represents the first demonstration for synthetic pollutants.

Additional and, at first glance, surprising results were presented by Möller et al. [19], who analysed the brain tissue of the same Eider ducks that had already been investigated by Kallenborn et al. with regard to liver, kidney, and muscle tissues. It turned out that an additional enantioselective process that thus far had escaped the attention of ecotoxicologists should be taken into account when assessing the potential risk of environmental pollutants: the enantioselective permeation through the blood-brain barrier. This notable effect will be discussed separately in Section 3.2.2.

Since the pioneering investigation by Kallenborn et al., who demonstrated the enantioselective metabolisation of α-HCH in common Eider ducks (*Somateria mollissima* L.) [17], increasing attention has been paid to the chromatographic enantiomer separation of chiral xenobiotics and their metabolites in environmental biota samples. A comprehensive survey of this rapidly developing field is hardly possible. Table 3.7 lists chiral environmental pollutants, as well as the relevant literature, for which a successful enantiomer separation of the respective standard compounds has been reported. Table 3.8 summarises environmental pollutants together with the respective biota samples and the references in which results about these compounds with regard to enantioselective analyses and processes can be found. The two lists represent the current status at the dead-line of the present volume, and it has to be supplemented continuously. In

Table 3.7. Standard compounds of environmental chiral xenobiotics for which successful enantiomer separations have been reported in the literature to date

Substance	Standards [Ref(s)]
α-HCH	standards [19–22]
β-PCCH	
γ-PCCH	
PCCH II, III, V-VIII	standards [21]
heptachlor	standards [22, 23, 25]
cis-chlordane	standards [20, 22, 23, 25]
trans-chlordane	standards [20, 22, 23, 25]
chlordene	standards [23]
heptachlor *exo*-epoxide	standards [20, 22, 23, 25]
heptachlor *endo*epoxide	standards [20, 22, 24, 25]
oxychlordane	standards [20, 22, 23, 25]
octachlordanes MC4, MC5, MC7, nonachlor III MC6	
photo-*cis*-chlordanes [2,5-, and 1,5-]	standards [22]
photoheptachlor	standard [16, 22]
bromocyclen	standards [20, 25]
o,p'-DDT	standards [22, 26]
o,p'-DDD	standards [22]
Hepta-, octa-, nonachlorobornanes	
toxaphene Parlar26/32/40/41/44/62	standards [27, 28]
toxaphene Parlar38/39/42/51/58/63	standards [28]
toxaphene Parlar50	standards [27]
PCB 45/139	standard [20, 29]
PCB 84/183	standards [20, 30]
PCB 91/136	standard [20, 31]
PCB 95	standard [20, 29, 31]
PCB 88/131/135/175	standards [20]
PCB 132	standard [20, 30–32]
PCB 144	standard [33]
PCB 149	standard [20, 30, 31]
PCB 171	standard [30]
PCB 174	standard [20, 30, 34]
PCB 176	standard [20, 34]
PCB 196/197	not yet separated
PCB methyl sulfones: 3–91, 4–91, 3–95, 4–95, 3–149, 4–149, 3–132, 4–132, 3–174, 4–174	standards [35, 36]
DCPP/MCPP	standards [23]
2-(2,4,5-trichlorophenoxy)propanoic acid	standard [23]
(*S*)-bioallethrin	standard [20]
methamidophos	standard [20, 23]
acephate	standard [20]
trichlofon	standard [20, 23]
malaoxon	standard [20]
bromoacil	standard [20]

Table 3.7. (continued)

Substance	Standards [Ref(s)]
fonofos	standard [23]
HHCB (Galaxolide®)/AHTN (Tonalide®)	standards [37, 111–113, 134]
ATII (Traseolide®)/AHDI (Phantolide®)	standards [37, 111–113]
galaxolidone	standard [134, 173]
(2R,2'R/S)-bis(2,3-dichloro-1-propyl) ether	standard [60]

Table 3.8. Environmental chiral or prochiral xenobiotics reported in the literature to date

Substance	Water/Tissue/Air [Ref(s)]
α-HCH	water [1, 2, 8, 9, 11–13, 15, 16, 38, 39, 58, 59]; common Eider duck: liver, muscle, kidney, brain [17, 38, 39]; seal: blubber, brain, liver, lung [19, 39–42, 56]; fish oil [41, 56, 57]; fish liver [38, 41]; blue mussel [38, 39]; whale blubber [42]; seagull eggs [39]; polar bear: liver, fat [56]; roe deer: liver [43]; sheep: fat, liver, brain [44]; air/rain [11–13, 15, 42, 58, 59]
γ-HCH	water [13, 22, 41, 42]
β-PCCH	water [1, 2, 21, 39]; air [2]
γ-PCCH	water [1, 2, 21]; air [2]
cis-chlordane	water [16]; herring, salmon, seal, penguin [24, 45]; cod liver oil [33]
trans-chlordane	water [16]; herring, salmon, seal, penguin [24, 45]; cod liver oil [33]
heptachlor exo-epoxide	water [16]; seagull eggs [25, 39]; herring, seal [24, 25, 46]; cod liver oil [33]; roe deer liver [25, 43]; hare liver [25]; rat liver [47]; human adipose [24]
oxychlordane	seagull eggs [25, 39]; herring, salmon [24]; cod liver oil [33]; hare liver [25]; roe deer liver [25, 43]; seal [5, 24, 25]; human adipose [24];
octachlordanes MC4, MC5, MC7	herring, salmon, seal, penguin [24, 33, 45]
nonachlor III MC6	herring, salmon, seal, penguin [24, 33, 45]
photo-cis-chlordanes [2,5-, and 1,5-]	herring, seal, salmon, penguin [24];
photoheptachlor	herring, salmon, penguin [24]; seal [24, 48]; polar bear, human plasma [48];
bromocyclen	water [49]; fish [49, 50];
hepta-, octa-, nonachlorobornanes	herring, salmon, seal, penguin [51]
PCB 88	blue mussel [52]
PCB 95	human milk, doe liver, eel [53]
PCB 132	doe liver, eel [32]; human milk [53, 54]

Table 3.8. (continued)

Substance	Water/Tissue/Air [Ref(s)]
PCB 149	blue mussel [52]; human milk, doe liver, eel [53]
PCB 171	blue mussel [52]
PCB 174	blue mussel [52]
PCB 183	blue mussel [52]
PCB methyl sulfones: 3–132, 3–149	human liver [35, 36, 172]; rats: liver, lung [104, 172]
2-(2,4-dichlorophenoxy)propanoic acid	water [6, 39]
2-(4-chloro-2-methylphenoxy)propanoic acid	water [5]
HHCB (Galaxolide®)	fish [37, 111–113, 134]
AHTN (Tonalide®)	fish [37, 111–113, 134]
ATII (Traseolide®)	fish [37, 111–113]

this present chapter, the discussion will be structured according to the chemical classes of the environmental pollutants. In our view, this will facilitate the comparison of enantioselective processes encountered at different trophic levels for the respective class of substances.

The chiral xenobiotic that has been investigated most intensively in all environmental compartments is, without a doubt, α-HCH. Its chemical structure, as well as its main microbial metabolism, has already been described in Section 3.1.1 (see also Fig. 3.1). In this section, emphasis is placed on enzymatic transformation of α-HCH in biota. Additional information about this compound can be found in Sections 3.2.2 (enantioselective permeation through the blood-brain barrier), 3.4 (photochemical conversion processes), 3.5.2 (chiral xenobiotics in soil), and 3.6 (air/water gas exchange and atmospheric long-range transport).

Already at lower trophic levels (microbial transformation in water, mussel), preferential depletion of one α-HCH enantiomer, be it the (+)- or the (–)-enantiomer, has been verified. These first results encouraged the implementation of a more systematic investigation that addressed the problem as to whether or not different enzymatic processes may also be revealed by the analysis of enantiomeric ratios of α-HCH in marine biota of different trophic levels. Therefore, during a cruise in 1991, Pfaffenberger et al. collected water samples, blue mussels, as well as flounders (*Platychthys flesus* L.) and common Eider ducks, in three areas typical of the German Bight, i.e., the mouth of the river, two coastal sites and in the central/north-western part of the German Bight [38]. All samples investigated were taken in marine areas that had been studied thoroughly since 1987 during several cruises of the projects ZISCH (Zirkulation und Schadstoffumsatz in der Nordsee) and PRISMA (Prozesse im Schadstoffkreislauf Meer-Atmosphäre: Ökosystem Deutsche Bucht) of the University of Hamburg. Thus, considerable background information about nutrients, heavy metals, and various organic pollutants was available, which could be used to define "more polluted" and "less polluted" sampling sites for the study carried out by Pfaffenberger et al.

On the basis of this background information, as a "more polluted" sampling site, the estuary of the River Elbe (F1, E1, M1, W1, W2; see Table 3.9) was chosen, while "less polluted" sampling sites include the estuary of the River Eider (F2),

Table 3.9. Concentrations and enantiomeric ratios of α-HCH for North Sea water, blue mussels, and for liver samples of common Eider ducks and of flounders, respectively; from [38]

Sample	Sampling site	Sample #	Concentration (ng/g wet weight)	(+)-α-HCH/ (−)-α-HCH	Mean enantiomeric ratio
North Sea water sample	W1	1	–	0.81	
	W2	2	–	0.84	
	W3	3	–	0.87	
	W4	4	–	0.83	0.84±0.03
Blue mussel	M1	1	0.93	0.70	
		2	0.78	0.97	
	M2	3	0.57	0.83	
		4	1.35	0.93	
		5	0.35	1.04	0.89±0.14
Common Eider duck (liver)	E1	1	3.12	2.8	
		2	0.21	∞	
		3	0.11	∞	
		4	1.67	1.8	
	E2	5	0.16	9.5	
		6	0.53	1.4	
		7	2.17	25	
		8	2.0	5.5	
Flounder (liver)	F1 (6/1991)	1	3.1	0.92	
		2	1.1	0.91	
		3	0.7	0.91	
		4	1.9	0.98	
		5	2.4	0.97	0.94±0.04
	F1 (1/1991)	6	1.1	0.83	
		7	1.0	0.84	
		8	2.1	0.76	
		9	2.0	0.76	0.80±0.05
	F2 (1/1991)	10	1.6	0.88	
		11	0.6	0.86	
		12	1.3	0.89	
		13	1.8	0.91	0.89±0.03

the Isle of Amrum (E2, M2) and the German Bight area off the River Elbe plume (W3, W4). The results thus obtained were compared with the aim of investigating the problem as to whether or not higher levels of pollutants may induce stronger enzymatic activities and thus larger shifts of the enantiomeric ratios of α-HCH. The mean enantiomeric ratio of α-HCH determined in the water samples, 0.84±0.03, compared well within the limits of error with earlier values for the eastern part of the North Sea and the Skagerrak (see Sect. 3.1.2). Furthermore, this value, which is assumed to represent the enantioselective transformation of α-HCH by marine microorganisms, seems to be independent of the seasonal variability in the biological activity.

The enantiomeric ratios determined for the blue mussel samples exhibit a larger variability, however, the values (mean ER value 0.89±0.14) largely reflect the fact that the mussels accumulate persistent organic pollutants like α-HCH, but cannot significantly degrade such persistent halogenated compounds. As a consequence, a similar enantiomeric ratio of α-HCH is observed to that found in the water which implies that at least no enantioselective transformation of this compound takes place in the blue mussel.

A dramatic shift of the enantiomeric ratio of α-HCH was, however, observed in the liver of the Eider ducks whose diet, as outlined above, almost completely consists of molluscs [163]. In all instances, the ratios (+)/(–)-α-HCH turned out to be significantly larger than 1, thus confirming the earlier results of Kallenborn et al. [17, 18]. A more detailed analysis in the study by Pfaffenberger included the aspects "health of the animals" as well as the contamination status of the sampling area [38].

Animals that had obviously been ill, e.g., two Eider ducks (samples #5 and #8; Table 3.9) had suffered from parasitic disease, showed enantiomeric α-HCH ratios between 1.4 (E2, sample #6) and 2.8 (E1, sample # 1), while Eider ducks in better condition exhibited enantiomeric α-HCH ratios up to nearly ∞ (E1, samples # 2 and #3), which implies that the (–)-α-HCH enantiomer had been almost completely decomposed, and only the (+)-α-HCH enantiomer was enriched in the liver. A comparison between samples of the more polluted site E1 and the less polluted site E2 shows no clear tendency, and the absolute concentrations of α-HCH in the liver samples (Table 3.9) also give no clear indication as to whether or not higher pollution of the area gives rise to stronger enzymatic activity. However, common Eider ducks must be considered as migrating marine birds, during spring and autumn moving long distances along the North Sea and Baltic coast. Thus, a direct correlation between local pollutant levels and the individual concentrations in the common Eider duck tissues is not to be expected. Moreover, in the case of the common Eider ducks, the ability to transform α-HCH in the liver appears to be rather dependent on the physical conditions of the animals, which, in turn, may be influenced by all kinds of diseases or by stress induced by marine pollutants. However, regardless of the absolute values of the enantiomeric ratios, in all instances the enzymatic processes in the liver of common Eider ducks gave rise to a faster (or nearly exclusive) degradation of (–)-α-HCH.

A different result was obtained from liver samples of the flounders. The mean values of the different test sites vary from 0.80 to 0.94 (Table 3.9), which reflect an enzymatic transformation process which prefers common structural elements represented by (+)-α-HCH. Since the enantiomeric ratios of α-HCH determined in the liver samples of the flounders in the two test sites F1 and F2 show relatively low variability for the respective areas, a comparison between these two data sets appears to be feasible: In the test site F1, a mean enantiomeric ratio of 0.80±0.05 was determined for the flounders caught in January 1991, while the flounders caught during the same period in the less polluted area F2 showed a mean value of 0.89±0.03. The result seems to support the assumption that a stronger enzymatic activity is induced in the liver of the flounders caught in the highly polluted Elbe estuary than in the liver of flounders living in the Eider estuary. However, it should be noted that the effects observed by Pfaffenberger et al. were based on a consistent but small data set, and, furthermore, the effect is only slightly beyond the limits of error. In addition, seasonal effects, e.g., different input of pollutants into the sea during winter and summer periods, may modify the results. This is demonstrated by a comparison of the enantiomeric ratios of α-HCH determined in flounders caught in the Elbe estuary in January 1991 (mean value of 0.80±0.05; F1/samples # 6–9, Table 3.9) and in June 1991 (mean value 0.94±0.04; F1/samples # 1–5). Therefore, caution has to be applied when interpreting the flounder data summarised in Table 3.9, and additional analytical data, which included systematic experiments with flounders in the laboratory, had to be awaited in order to verify the hypothesis that higher levels of pollutants may induce the enzymatic activity in the liver of flounders (Sect. 4.3.4).

Meanwhile, several authors have addressed the problem *enantioselective transformation of α-HCH* by marine animals at higher trophic levels, such as seals [19, 39–42, 92], whales [42], and polar bears [56], by applying chiral selectors on the basis of modified cyclodextrin phases. The results are summarised in Table 3.10. In detail, different emphasis was placed on the investigations on marine mammals published in [19, 39–42, 56, 92]. For example, Hühnerfuss et al. compared the results of tissue sample extracts of dead harbour seals (*Phoca vitulina* L.) from the German Bight and from Iceland, respectively, in order to study the impact of different contamination levels on the enantioselective processes [40]. In the seal tissues obtained from animals of the German Bight, enantiomeric ratios of 1.5 to 4.5 were determined for blubber tissue, while brain tissue from the same animals yielded enantiomeric ratios of 7.9 to 19.9. These values have to be compared with those from Iceland seals, i.e., enantiomeric ratios in blubber 1.2–1.4 and in brain tissues 55.6, 66.2 and six values of nearly ∞. Though the general tendency of the data from both regions is comparable, the enantiomeric ratios in the blubber of Iceland animals, i.e., from the less polluted area, appear to be lower, and in the brain samples from the eight Iceland seals almost exclusively (+)-α-HCH was found. As the α-HCH concentrations in blubber of the German Bight seals were about five times higher than those of the Iceland seals, it is tentatively assumed that the higher concentrations may have induced higher enzy-

Table 3.10. Enantiomeric ratios [(+)-/(–)-enantiomer] of α-HCH in extracts of marine biota tissues of higher trophic levels; for comparison, some cod samples (oil and whole body, respectively) are included (M = male, F = female)

Sample	Tissue	Enantiomeric ratios			Ref(s)
		Column A	Column B	Column C	
Neonatal fur seal (healthy)	blubber	1.88	1.67	–	[41]
(*Callorhinus ursinus* L.)	liver	1.66	1.45	1.64	
	lung	1.55	1.47	–	
	brain	30.0	26.2	26.2	
Stillborn fur seal	blubber	1.85	1.60	–	[41]
(*Callorhinus ursinus* L.)	liver	1.83	1.64	–	
	brain	32.9	31.4	–	
Neonatal fur seal (diseased)	blubber	1.20	1.18	–	[41]
	liver	1.30	1.27	–	
Female fur seal	milk	1.58	1.47	–	[41]
Cod liver oil North Atlantic	SRM 1588, NIST	0.98	0.99	–	[41]
Harbour seal (German Bight)	blubber	4.47	–	–	[40]
(*Phoca vitulina* L.)	spinal marrow	19.9	–	–	
Harbour seal (German Bight)	blubber	1.54	–	–	[40]
(*Phoca vitulina* L.)	brain	16.3	–	–	
Harbour seal (German Bight)	blubber	2.83	–	–	[40]
(*Phoca vitulina* L.)	brain	7.9	–	–	
Harbour seal (Iceland)	blubber	1.36	–	–	[19, 40]
(*Phoca vitulina* L.)	brain	∞	–	–	
	spinal marrow	∞	–	–	
Harbour seal (Iceland)	blubber	1.26	–	–	[19, 40]
(*Phoca vitulina* L.)	brain	∞	–	–	
	spinal marrow	99.5	–	–	
Harbour seal (Iceland)	blubber	1.21	–	–	[19, 40]
(*Phoca vitulina* L.)	brain	∞	–	–	
	spinal marrow	∞	–	–	
Harbour seal (Iceland)	brain	∞	–	–	[19, 40]
(*Phoca vitulina* L.)	spinal marrow	∞	–	–	
Harbour seal (Iceland)	brain	66.2	–	–	[19, 40]
(*Phoca vitulina* L.)	spinal marrow	30.2	–	–	
Harbour seal (Iceland)	brain	55.6	–	–	[19, 40]
(*Phoca vitulina* L.)	spinal marrow	∞	–	–	
Harbour seal (Iceland)	brain	∞	–	–	[19, 40]
(*Phoca vitulina* L.)	spinal marrow	∞	–	–	
Harbour seal (Iceland)	brain	∞	–	–	[19, 40]
(*Phoca vitulina* L.)	spinal marrow	22.8	–	–	
Harbour seal	blubber	2.33	2.04	–	[42]
(*Phoca vitulina* L.)					

Table 3.10. (continued)

Sample		Tissue	Enantiomeric ratios			Ref(s)
			Column A	Column B	Column C	
Harbour seal (*Phoca vitulina* L.)		blubber	2.17	2.17	–	[42]
Whale blubber (*Ph. Phoceona* L.)		blubber	2.78	2.51	–	[42]
Whale blubber (*Ph. Phoceona* L.)		blubber	2.86	2.63	–	[42]
Arctic Cod (*Boreogadus saida* LEP.)		whole body	≈1	–	–	[56]
Ringed seal (*Phoca hispida* ERX)		blubber	≈1	–	–	[56]
		liver	≈1.4	–	–	
Polar bear (*Ursus maritimus* PH)		fat	≈1.7	–	–	[56]
		liver	≈2.5	–	–	
Harbour seal (*Phoca vitulina* L.)	F	blubber	1.0			[92]
	F	blubber	1.3			
	F	blubber	1.3			
	F	blubber	1.3			
	F	blubber	1.8			
	M	blubber	1.2			
	M	blubber	1.3			
	M	blubber	1.5			
		Mean Value	1.4			
Grey seal (*Halichoerus grypus* FABR.)	F	blubber	1.0			[92]
	F	blubber	1.3			
	F	blubber	1.2			
	F	blubber	1.2			
	F	blubber	1.5			
	M	blubber	1.5			
	M	blubber	1.5			
	M	blubber	1.2			
		Mean Value	1.3			
Cod liver oil (*Gadus morhua* L.)			1.00–1.23			[57]
Fish oil			1.0			

matic activity, which in turn is expected to give rise to higher enantiomeric shifts in blubber. The aspect of enantioselective permeation of α-HCH enantiomers through the blood-brain barrier, which can also be inferred from the data set thus far available, will be discussed separately in Section 3.2.2.

Mössner et al. [41] focused on the enantioselective transformation of α-HCH in ill and healthy neonatal as well as adult fur seals (*Callorhinus ursinus* L.). Tissues of three dead neonatal fur seals with different health status were collected on St. Paul Island (Pribilof Islands, Alaska, USA) in July 1990 in cooperation with the Alaska Marine Mammal Tissue Archival Project and the National Marine Fisheries Service. The animals were characterised as follows: firstly, a neonatal fur seal (healthy): death due to massive head trauma; body condition good; age 1–2 weeks; male, secondly, a stillborn fur seal; female; thirdly, a neonatal fur seal (diseased); white muscle syndrome; body condition good; age 7–10 days; male. Furthermore, a milk sample, which was taken from the stomach of a fur seal pup that had died of pneumonia, was included in the study by Mössner et al. [41]. A cod liver oil sample, originating from the North Atlantic, reflecting in part the sum of abiotic and biotic transformation processes of the diet, was analysed as a reference (SRM 1588, NIST, Gaithersburg, Maryland, USA).

For blubber, liver, lung, and milk tissues of the fur seals, Mössner et al. obtained enantiomeric ratios of α-HCH in the range 1.2 to 1.9 (Table 3.10), i.e., values which are in accordance with those reported by Hühnerfuss et al. for the Iceland seal blubber tissues, and, for the two brain tissues, high ratios of 28 and 32 were also determined. The ratio found in the milk sample reflects enantioselective processes in the mother animal. The slightly enhanced ratios in the tissues of the normal and stillborn fur seals are assumed to be indicative of additional metabolic discrimination of the later-eluting enantiomer. However, the neonatal fur seal, which died of white muscle syndrome, shows the smallest ratios. It cannot be excluded that the disease may have influenced the enzyme induction potential and thus reduced the potential to metabolise xenobiotics like α-HCH.

Müller et al. [92] extended the data sets thus far available for the enantiomeric composition of α-HCH in seal tissues by a systematic investigation including two different seal species, eight harbour seals (*Phoca vitulina* L.) and eight grey seals (*Halichoerus grypus* FABR.), which had been shot on western Iceland. For both seal species, ER values of $ER_{+/-}$ >1 were determined in blubber extracts, which is fully in accordance with the results of other authors (Table 3.10). With mean ER values of 1.4 (harbour seals) and 1.3 (grey seals), no species-dependent effects were observed for these two seal species. Up to now, hooded seals (*Cystofora cristata* L.) are the only marine mammals for which an ER <1 for α-HCH have been reported [97]. Müller et al. did not verify any correlation between α-HCH levels or enantiomeric ratios with age or sex for the two seal species.

Wiberg et al. [56] investigated enantioselective processes of organochlorines, including α-HCH in the Arctic marine food chain, placing special emphasis on the polar bear (*Ursus maritimus* PH.) food chain, which is simple due to the limited biodiversity in the arctic marine environment. The main food for polar bears is the blubber of ringed seals (*Phoca hispida*, ERX.), which in turn largely consume Arctic cod (*Boregadus saida* LEP.) and amphipods.

The Arctic cod tissue extracts showed near-racemic ERs, indicating that bioaccumulation takes place without or with minor selective metabolism. This conclusion is fully in accordance with results of other authors who have reported enantiomeric ratios of α-HCH in cod liver or fish oil extracts of between 0.98 and 1.23 [41, 57]. The tissue extracts of ringed seals exhibited large changes in ER. Obviously, (+)-α-HCH becomes more abundant relative to (−)-α-HCH in top predators (Fig. 3.12). Therefore, Wiberg et al. calculated separate biomagni-

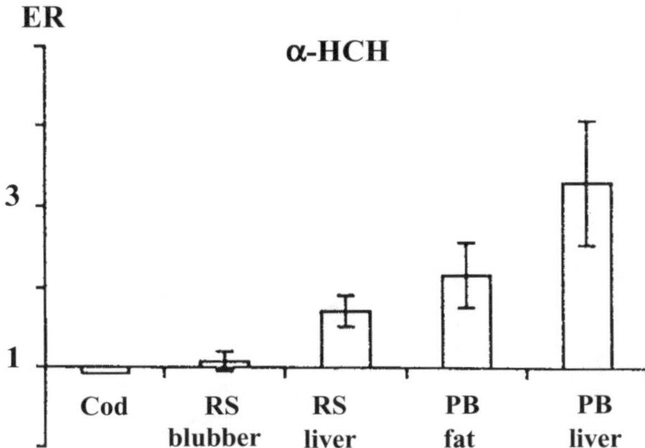

Fig. 3.12. Average enantiomeric ratios [ER; (+)-/(−)-enantiomer] ± standard deviation of α-HCH in the polar bear (PB) food chain; RS=ringed seal; from [56]

fication factors (BMFs) for the (+)- and the (−)-enantiomers. For the first step, from Arctic cod to seal, the BMFs are comparable [(+)-α-HCH: BMF=2.0; (−)-α-HCH: BMF=1.7]; however, during the next step, from seal to polar bear, only the (+)-enantiomer biomagnifies (BMF=1.4), in contrast to the (−)-enantiomer (BMF=0.7). For the complete chain, from cod to polar bear, the values are for the (+)-enantiomer BMF=2.8 and for the (−)-enantiomer BMF=1.2, respectively. The data set used for the calculations by Wiberg et al. comprised 40 samples. At least qualitatively, the results are fully in line with the observation that the (+)-enantiomer is increasingly dominant up the polar bear food chain.

Several persistent polychlorinated insecticides are chiral and have been applied as racemates. More than 15 years after the ban on their application in the United States, Japan and most European countries, they are still found in marine and terrestrial ecosystems at all trophic levels. This is true, in particular, for cyclodiene pesticides, e.g., *cis*- and *trans*-chlordane, heptachlor, aldrin and dieldrin, as well as some of their oxygenated metabolites like oxychlordane and heptachlor *exo*-epoxide. These compounds were extensively used in some countries (e.g., U.S.A.), but reportedly not in others (e.g., in Scandinavia). The fact that they are still being detected in biota from remote areas, for example, the Arctic and Antarctic, points to global distribution mechanisms such as long-range atmospheric transport. This latter aspect will be discussed in Section 3.6.

In 1991, König et al. were the first to report the successful enantiomer separation of *cis*- and *trans*-chlordane, oxychlordane, heptachlor, and heptachlor *exo*-epoxide (see Fig. 3.13) using selectively derivatised cyclodextrin phases [23]. Although the authors were able to resolve the individual enantiomeric pairs of *cis*- and *trans*-chlordane on several cyclodextrin derivatives [e.g., heptakis(3-*O*-methyl-2,6-di-*O*-*n*-pentyl)-β-cyclodextrin or per-*O*-methyl-β-cyclodextrin], the separation of all four stereoisomers was possible only on heptakis(2-*O*-methyl-3,6-di-*O*-*n*-pentyl)-β-cyclodextrin.

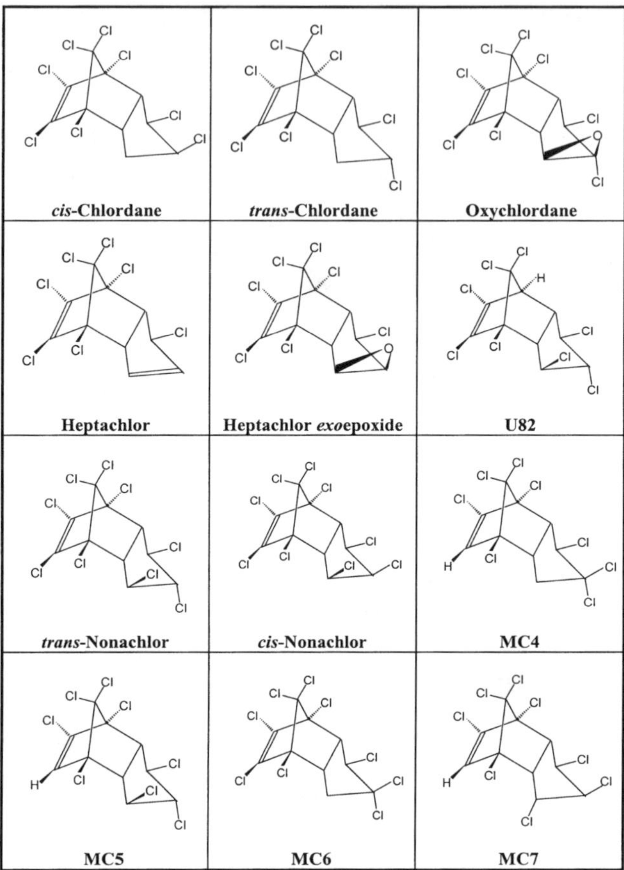

Fig. 3.13. Molecular structures of *cis*- and *trans*-chlordane, oxychlordane, heptachlor, heptachlor *exo*-epoxide, the chiral chlordanes U82, MC4, MC5, MC6, MC7, and the achiral *trans*- and *cis*-nonachlor (derived in part from [61])

Meanwhile, several authors have applied this new experimental approach to the investigation of enzymatic transformation of cyclodiene pesticides and their metabolites in marine and limnic biota (see Tables 3.11 and 3.12; [24, 25, 33, 39, 45, 46, 48–50, 57, 64–66, 92]). Buser et al. [45] analysed herring oil and tissues of a salmon, a seal and a penguin. The herring oil was prepared from fresh herring (*Clupea harengus* L.) collected from the Gulf of Bothnia during June 1988. Preparation of the oil was performed at the Norwegian Herring Oil and Meal Industry Research Institute, Bergen, Norway. The salmon muscle tissue stemmed from a female salmon (*Salmo salar* L.) caught in the River Ume at Stornorrfors, Sweden, in 1989. The seal sample was a composite of liver tissue of adult grey seal (*Halichoerus grypus* FABR.) collected from the Baltic Sea along the Swedish south-eastern coastline during the 1980s. Finally, the penguin tissue was a sample from a juvenile Adelie penguin (*Pygoscelis adeliae* L.) found dead at Shack-

Table 3.11. Enantiomeric ratios of *cis*- (CC) and *trans*-chlordane (TC), oxychlordane (OXY), heptachlor (HEPTA), heptachlor *exo*-epoxide (HEPX) determined in tissue extracts of marine and limnic biota; F=female (age in years); M=male (age in years)

Biota/Tissue			Enantiomeric ratios					Ref(s)
			CC	TC	OXY	HEPTA	HEPX	
Baltic herring			1.35±0.1	0.42±0.02	–	–	–	[24, 45]
Baltic salmon			0.38±0.03	1.19±0.02	–	–	–	[24, 45]
Baltic seal			–	0.60±0.04	–	–	–	[24, 45]
Antarctic penguin			–	–	–	–	–	[24, 45]
Harbour seal: liver			–	–	0.57	–	0.06	[25, 64]
blubber			–	–	0.66	–	0.14	
Harbour seal: liver			–	–	0.45	–	0.05	[25, 64]
blubber			–	–	0.54	–	0.18	
Seagull egg			–	–	2.3		–	[25, 64]
Seagull egg			–	–	2.1		–	[25, 64]
Seagull egg			–	–	1.8		–	[25, 64]
Seagull egg			–	–	1.5		2.7	[25, 64]
Seagull egg			–	–	1.5		1.6	[25, 64]
Cod liver oil NIST			0.95	1.13	1.33	–	1.60	[33]
Cod:	muscle	F6	0.71	0.54	–	–	–	[65]
	gonad		0.77	0.68	–	–	–	
	liver		0.71	0.54	–	–	–	
Cod:	muscle	F12	0.83	0.73	–	–	–	[65]
	gonad		1.0	0.77	–	–	–	
	liver		0.83	0.62	–	–	–	
Cod:	muscle	F11	0.91	0.46	–	–	–	[65]
	gonad		1.1	0.62	–	–	–	
	liver		0.91	0.41	–	–	–	
Cod:	muscle	M9	1.14	1.9	–	–	–	[65]
	gonad		1.1	2.0	–	–	–	
	liver		1.2	2.2	–	–	–	
Cod:	muscle	M6	1.0	1.1	–	–	–	[65]
	gonad		1.0	1.1	–	–	–	
	liver		1.06	1.2	–	–	–	
Herring		F3	1.1	0.48	–	–	–	[65]
		M3	1.2	0.20	–	–	–	
		M3	1.2	0.27	–	–	–	
		M3	1.3	0.58	–	–	–	
		M3	1.5	0.47	–	–	–	
		F3	1.2	0.36	–	–	–	
		F3	1.2	0.35	–	–	–	
		F3	1.6	0.27	–	–	–	
		F3	1.6	0.32	–	–	–	
Cod liver oil			0.50–1.02	0.80–3.49	–	–	–	[57]
Fish oil			1.00–1.18	1.00–1.06	–	–	–	[57]
Harbour seal:		F	–	–	0.7	–	–	[92]
blubber		F	–	–	0.7	–	–	

Table 3.11. (continued)

Biota/Tissue		Enantiomeric ratios					Ref(s)
		CC	TC	OXY	HEPTA	HEPX	
	F	–	–	0.8	–	–	
	F	–	–	0.6	–	–	
	F	–	–	0.7	–	–	
	M	–	–	0.7	–	–	
	M	–	–	0.6	–	–	
	M	–	–	0.9	–	–	
Mean Value		–	–	0.7	–	–	
Grey seal	F	–	–	1.6	–	–	[92]
blubber	F	–	–	1.5	–	–	
	F	–	–	1.4	–	–	
	F	–	–	1.2	–	–	
	F	–	–	1.1	–	–	
	M	–	–	1.1	–	–	
	M	–	–	1.4	–	–	
	M	–	–	1.5	–	–	
Mean Value		–	–	1.3	–	–	

leton's Hut, Ross Island, Antarctica, in March 1988. The frozen tissue was shipped to the University of Umeå, Sweden, where extraction and clean-up were carried out (for details of the experimental procedure, see [45]).

The authors stress that no individual chlordane congeners were available as standards for their investigation. Instead, they used the technical chlordane mixture, in which all compounds of interest could be identified and then assigned in the environmental samples using published retention data. Technical chlordane consists of a complex mixture of primarily hepta-, octa-, and nonachlorinated tricyclic compounds. It is prepared by the Diels-Alder reaction of hexachlorocyclopentadiene and cyclopentadiene to chlordene, an unsaturated hexachlorinated, tricyclic compound. Subsequent chlorination leads to products containing two to three additional chlorine atoms in the cyclopentane ring, i.e., octa- and nonachlordanes. The main constituents in technical chlordane, *cis*- and *trans*-chlordane and *trans*-nonachlor, belong to these groups. Furthermore, there are a number of more complex structures present in the technical mixture resulting from rearrangement reactions, in particular, Wagner-Meerwein rearrangements [62]. Figure 3.13 shows the molecular structures of *cis*- and *trans*-chlordane, oxychlordane, heptachlor, heptachlor *exo*-epoxide, U82, MC4, MC5, MC6, MC7 and *trans*- and *cis*-nonachlor. All derivatives possess an *endo* configuration, which can be easily explained on the basis of the frontier orbital theory valid for pericyclic reactions like the Diels-Alder reaction [63]. Furthermore, all compounds shown in Fig. 3.13 are chiral, apart from *cis*- and *trans*-nonachlor.

The chlordane components detectable in all four biota samples investigated by Buser et al. [45] were the chiral congeners *cis*- and *trans*-chlordane, MC4, MC5, MC6, MC7, and U82, the prochiral *cis*- and *trans*-nonachlor, as well as var-

Table 3.12. Enantiomeric ratios of additional cyclodiene pesticides, in particular, chiral chlordane components observed in technical chlordane mixtures and bromocyclen (BC), determined in tissue extracts of marine and limnic biota; F = female (age in years); M = male (age in years)

Biota/Tissue	Enantiomeric ratios								Ref(s)
	MC4	MC5	MC6	MC7	K	U81	U82	BC (−)/(+)	
Baltic herring	0.7	0.81	–	0.83	–	–	–	–	[24, 45]
Baltic salmon	0.7	0.75	–	0.92	–	–	–	–	[24, 45]
Baltic seal	2.7	0.24	–	0.86	–	–	–	–	[24, 45]
Antarctic penguin	2.2	0.91	–	1.35	–	–	–	–	[24, 45]
Cod liver oil NIST	–	0.87	–	–	–	–	not res.	–	[33]
Cod: muscle F6	–	–	0.29	–	–	0.85	0.73	–	[65]
gonad	–	–	0.23	–	–	1.0	0.76	–	
liver	–	–	0.11	–	–	0.93	0.76	–	
Cod: muscle F12	–	–	0.32	–	–	0.77	0.80	–	[65]
gonad	–	–	0.32	–	–	0.76	0.75	–	
liver	–	–	0.26	–	–	0.88	0.76	–	
Cod: muscle F11	–	–	0.54	–	–	0.83	0.82	–	[65]
gonad	–	–	0.35	–	–	0.85	0.74	–	
liver	–	–	0.24	–	–	0.78	0.76	–	
Cod: muscle M9	–	–	1.6	–	–	0.93	0.97	–	[65]
gonad	–	–	1.0	–	–	0.92	0.76	–	
liver	–	–	1.1	–	–	0.95	0.79	–	
Cod: muscle M6	–	–	1.8	–	–	–	0.92	–	[65]
gonad	–	–	1.1	–	–	1.1	0.89	–	
liver	–	–	1.0	–	–	0.86	0.89	–	
Herring F3	–	–	–	–	–	–	1.1	–	[65]
M3	–	–	–	–	–	–	1.2	–	
M3	–	–	–	–	–	–	1.2	–	
M3	–	–	–	–	–	–	1.1	–	
M3	–	–	–	–	–	–	1.2	–	
F3	–	–	–	–	–	–	1.0	–	
F3	–	–	–	–	–	–	1.0	–	
F3	–	–	–	–	–	–	1.1	–	
F3	–	–	–	–	–	–	1.0	–	
Trout	–	–	–	–	–	–	–	0.84	[50]
	–	–	–	–	–	–	–	0.93	
	–	–	–	–	–	–	–	1.00	
	–	–	–	–	–	–	–	0.94	
	–	–	–	–	–	–	–	0.84	
Orfe	–	–	–	–	–	–	–	0.80	[50]
Bream	–	–	–	–	–	–	–	0.89	[50]
Pike	–	–	–	–	–	–	–	0.85	[50]
Trout: muscle	–	–	–	–	–	–	–	0.66	[49, 66]
	–	–	–	–	–	–	–	0.71	

Table 3.12. (continued)

Biota/Tissue	Enantiomeric ratios								Ref(s)
	MC4	MC5	MC6	MC7	K	U81	U82	BC (–)/(+)	
bream: muscle	–	–	–	–	–	–	–	1.19	[49, 66]
	–	–	–	–	–	–	–	1.22	
	–	–	–	–	–	–	–	1.08	
	–	–	–	–	–	–	–	1.34	
	–	–	–	–	–	–	–	1.56	
	–	–	–	–	–	–	–	2.13	
	–	–	–	–	–	–	–	1.93	
	–	–	–	–	–	–	–	2.05	

ious minor components. Component K was not found to be present in these extracts. For some congeners, a complete enantiomer separation was achieved by Buser et al., for other components enantiomer separation remained incomplete (see Tables 3.11 and 3.12). The most dissimilar enantiomeric composition (ER value of 0.24) was observed for MC5 in Baltic seal. The major octachlordane component in both fish samples was *cis*-chlordane, whereas in herring the earlier- and in salmon the later-eluting enantiomer predominated, i.e., the enantiomeric ratios of *cis*-chlordane between the two fish species are thus reversed. For *trans*-chlordane, the authors also observed a reversal in the enantiomeric ratios between the two species. In this case, the later-eluting enantiomer predominated in herring and the earlier-eluting in salmon.

In seal and penguin, the concentrations of *cis*- and *trans*-chlordane were much lower and, therefore, enantiomeric ratios were much more difficult to determine. In these two warm-blooded species, the components U82 and MC5 were dominant. U82 is a 5+3-type chiral octachlordane of, at that time, unknown configuration [24]. At first, Buser et al. were not able to resolve the enantiomers of U82, but their endeavour was successful in a subsequent investigation, when they applied a different chiral selector, heptakis(6-*O*-*tert*-butyl-dimethylsilyl-2,3-di-*O*-methyl)-*β*-cyclodextrin (TBDMS-CD), to a technical chlordane mixture [24]. Furthermore, Karlsson et al. meanwhile successfully elucidated the structure of U82 (see Fig. 3.13) [166, 167]. By contrast, MC5 was clearly separated into enantiomers (Table 3.12). The later-eluting enantiomer predominated in seal tissue, and some preference for this enantiomer was still observed in tissues of the other aquatic species included in the study. Some differences in the enantiomeric composition of components MC4 and MC7 were also inferred from the data set. For component MC4, the chromatograms showed a clear predominance of the earlier-eluting enantiomer in the warm-blooded species whereas this can hardly be observed in the two fish samples. For component MC7, the chromatograms indicated a small preference for the later-eluting enantiomer in herring, salmon, and seal, whereas in the penguin a small prevalence of the earlier-eluting enantiomer can be recognised.

Some of the separation problems encountered by Buser et al. with regard to components of the technical chlordane mixture were overcome by Karlsson et al., who used heptakis(2,3,6-*O-tert*-butyldimethylsilyl)-β-cyclodextrin as chiral selector. These latter authors analysed Atlantic cod (*Gadus morhua* L.) samples carrying out a very comprehensive study. Atlantic cod, which was moving from the Barent Sea to the Lofot Islands for spawning, was caught outside Kvaløya, Troms County, Norway (70°N, 17°E). Age, sex, size, as well as maturation stage of the gonads, were determined. Liver and gonads were weighed and samples were taken including a portion of the fillet. The extraction and clean-up procedure can be found in [65].

As can be seen from Tables 3.11 and 3.12, the enantiomeric ratios (ER; area of (+)/(–)-enantiomer or first-eluting enantiomer divided by the second one) in Atlantic cod deviate substantially from the racemic ratio. With a few exceptions, the ERs found in all three cod tissues are similar within the uncertainty of the method (3–5%). This means that any tissue can be used for an ER determination. It is worth noting that the ER of the octachloro congener U82 was below one in cod and above one in herring. No differences in the ERs between male and female herring were observed. However, in all cod samples analysed by Karlsson et al., the ERs for *trans*-chlordane and MC6 were very different between males and females. Enantiomer degradation was opposite in male and female cod leading to changes in the ERs by a factor of 3–5. A similar but not so unequivocal trend was also obtained for *cis*-chlordane. No gender difference was observed for U81 and U82. The differences in influence of sex on the ERs of the congeners cannot be explained yet. Karlsson et al. [65] conjecture that U81 and U82 have a 5+3 chlorine distribution between ring 2 and 1, while all other congeners are of the 6+2 (octachloro congeners) or 6+3 type (nonachloro congeners). The technical pesticide chlordane has shown a significant synergism concerning artificial estrogenic effects together with other pesticides such as dieldrin [67]. This might be an explanation for the different ERs in male and female cod. However, this does not explain the missing gender influence in herring. Other factors, such as position in the food chain and unlike enzyme systems, can also be of importance.

The enantioselective analysis of organochlorines in the Arctic food chain "Arctic cod/seal/polar bear" carried out by Wiberg et al. (see above [56]) also included the chlordane congeners MC4, MC5, and MC6. The results are summarised in Fig. 3.14. Similar to the situation encountered in the case of α-HCH, near-racemic ERs were found in the cod tissue extracts. With regard to the octachlordane MC5, the second-eluting enantiomer became more abundant up the food chain. Different enantiomeric prevalence was observed for MC4: while the first-eluting enantiomer dominated in blubber and liver of the ringed seal, the second-eluting enantiomer was preferentially found in the liver tissue extract of the polar bear. In the case of MC6, both in the ringed seal and polar bear tissues, the second-eluting enantiomer dominated. A strange reversal of the enantiomers up the food chain was determined for the oxygenated metabolites. The (+)-enantiomer of heptachlor *exo*-epoxide was in excess in Arctic cod and polar

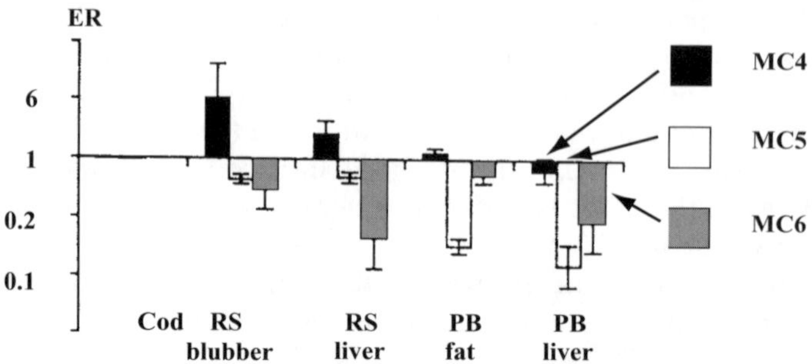

Fig. 3.14. Average enantiomeric ratios (ER)±standard deviation of the chlordane congeners MC4, MC5, and MC6 in the polar bear (PB) food chain; RS=ringed seal; from [56]

bear, whereas the (–)-enantiomer dominated in ringed seals. The mean ER for oxychlordane was similar for all species in the food chain.

As already explained for α-HCH, Wiberg et al. also calculated separate biomagnification factors (BMFs) for the (+)- and the (–)-enantiomers of the oxygenated compounds. For the first step, from Arctic cod to seal, the BMFs are comparable for oxychlordane [(+)-OXY: BMF=152; (–)-OXY: BMF=124], and also for the second step, from seal to polar bear, similar values were calculated [(+)-OXY: 7.1; (–)-OXY: 6.7]. As a consequence, the overall values from Arctic cod to polar bear are also comparable [(+)-OXY: 1075; (–)-OXY: 834]. The strange reversal of the enantiomeric ratios observed for heptachlor *exo*-epoxide up the food chain is reflected by the following BMFs: from Arctic cod to seal for (+)-HEPX BMF=4.8 and for (–)-HEPX BMF=9.3; from seal to bear for (+)-HEPX BMF=7.4 and for (–)-HEPX BMF=2.3; from cod to bear for (+)-HEPX BMF=35 and for (–)-HEPX BMF=21.

After the first successful separation of the enantiomers of oxychlordane and heptachlor *exo*-epoxide [23], the main metabolites of *cis-/trans*-chlordane and heptachlor, respectively, Hühnerfuss et al. performed systematic investigations on the enantiomeric distribution of these metabolites in various marine biota samples [25, 64]. Oxychlordane was found in all five sea-gull eggs investigated, while the concentrations of heptachlor *exo*-epoxide were below the detection limit in three eggs. The presence of heptachlor *exo*-epoxide shows that the transformation of both *exo*heptachlor and *endo*heptachlor to heptachlor *exo*-epoxide is not confined to mammalian biota like seals, but it is also the exclusive transformation pathway in sea birds. The ER values of oxychlordane and heptachlor *exo*-epoxide in sea-gull eggs are of particular interest: in both cases, the values were larger than one, indicating a preferential accumulation of the respective (+)-enantiomer. In contrast, the corresponding values determined in different tissues of seals were smaller than one. This may possibly suggest that sea-gulls reflect rather terrestrial characteristics than marine ones (cf. Sect. 3.2.1.2; Tables 3.16 and 3.17). This holds at least for the enzymatic processes that give rise to enantioselective accumulation of oxychlordane and/or heptachlor *exo*-epoxide. The results obtained for the seal tissues will be further discussed in

Section 3.2.2 in connection with the permeation of the cyclodiene metabolites through the blood-brain barrier.

Oxychlordane enantiomers were also analysed by Müller et al. [92] in harbour seals (*Phoca vitulina* L.) and grey seals (*Halichoerus grypus* FABR.). They reported the notable result that higher (–)-oxychlordane levels were present in harbour seals, while in grey seals (+)-oxychlordane dominated (Table 3.11). A more detailed analysis of the data set revealed that neither levels nor enantiomeric ratios of α-HCH and oxychlordane did correlate in harbour seals or grey seals. In addition, no trend in the ER values of oxychlordane with regard to age or sex could be inferred from the data obtained by Müller et al.

In the last few years, another cyclodiene pesticide, the insecticide bromocyclen, has drawn the attention of several ecochemists [49, 50, 66]. It possesses neither a plane nor a centre of symmetry and, therefore, it exists in two enantiomeric forms (see Sect. 3.1.2; Fig. 3.10). Due to its very low mammalian toxicity, bromocyclen is used in Europe against ectoparasites for the treatment of domestic animals. In 1993, Rimkus and Wolf [175]were the first to report the contamination of both German and imported samples of rainbow trout with this new environmental pollutant. This was in itself surprising, as there is no official registration of this substance for fish-farming.

The appearance of this new environmental contaminant stimulated two investigations by Pfaffenberger et al. [50] and by Bethan et al. [49, 66]. Pfaffenberger et al. analysed fish samples from various fish farms in Denmark and from the River Stör, a tributary of the River Elbe in the northern German state Schleswig-Holstein. They focused on the problem as to whether a significant enzymatic transformation of bromocyclen is possible at that trophic level. Particular emphasis was placed on the question whether a correlation between the concentration of bromocyclen and its enantiomeric ratio can be inferred from enantioselective cGC using modified cyclodextrins as chiral selector.

The fish samples were collected during the spring of 1990 and the autumn of 1992. The rainbow trout (*Oncorhynchus mykiss* WAL.) were imported from various fish farms in Denmark. The other fish species, orfe (*Leuciscus idus* L.), bream (*Abramis brama orientalis* BERG) and pike (*Esox lucius* L.), were caught in the River Stör. Details of the sample preparation and clean-up can be found in [50]. In all eight fish samples in part remarkably high concentrations of bromocyclen were determined. They varied between 0.093 and 1.200 mg/kg fat, regardless of whether the fish lived in the artificial environment of a fish farm or in a natural surrounding like the River Stör. Furthermore, no correlation between the fish species and the observed concentrations was found.

With regard to the determination of enantiomeric excesses, Pfaffenberger et al. were the first to separate the enantiomers of bromocyclen in fish tissue extracts. An example of these first successful enantiomer separations with the help of a 25 m fused-silica capillary column, coated with 50% (w/w) heptakis(6-*O*-*tert*-butyldimethylsilyl-2,3-di-*O*-methyl)-β-cyclodextrin and 50% OV-1701, is shown in Fig. 3.15. The enantiomeric ratios as determined in the fish samples are summarised in Table 3.12. The ERs vary between 0.84 and 1.00. In all samples

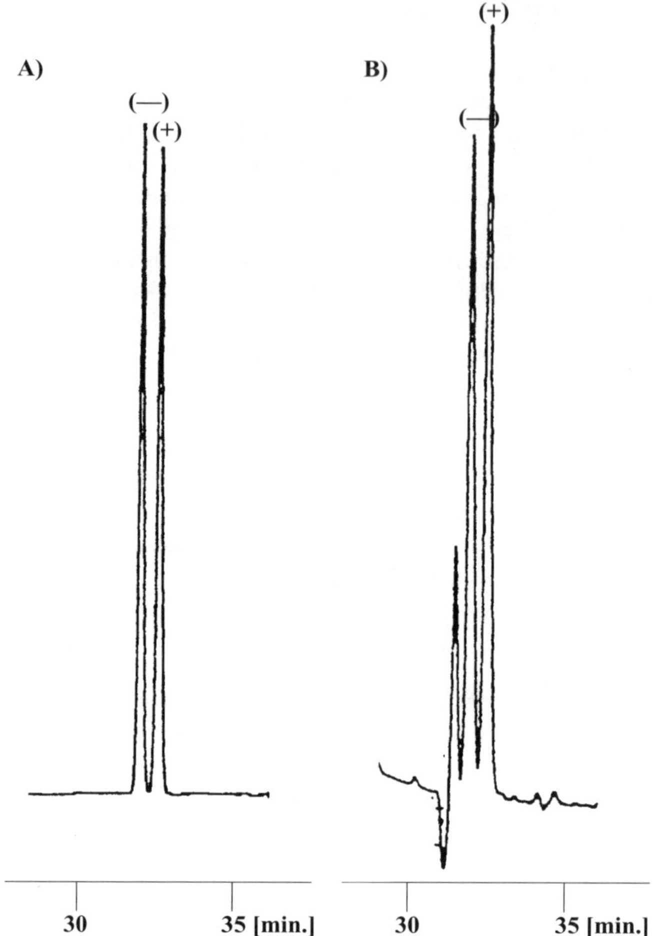

Fig. 3.15A, B. Enantiomer separation of bromocyclen extracted from fish samples. **A** Standard of technical (racemic) bromocyclen. **B** sample of rainbow trout from a Danish fish farm showing different enantiomeric ratios of bromocyclen. cGC conditions: fused-silica capillary column coated with 50% (w/w) heptakis(6-*O-tert*-butyldimethylsilyl-2,3-di-*O*-methyl)-β-cyclodextrin and 50% OV-1701. Column temperature program: initial 333 K, increased at 10 K/min to 413 K, 40 min isothermal; carrier gas hydrogen (60 kPa); on-column injection; ECD detection; from [50]

of the different fish species the first-eluting (–)-enantiomer was preferentially degraded, with the exception of one sample which showed an ER of 1.00 (the assignment of the order of elution of enantiomers is based on optical rotation measurements after enantiomeric resolution by preparative packed-column GC [25]). This effect was observed regardless of whether the sample originated from fish farms or from the River Stör. Therefore, it can be safely assumed that an enzymatic process, which is common to all fish species investigated by Pfaffen-

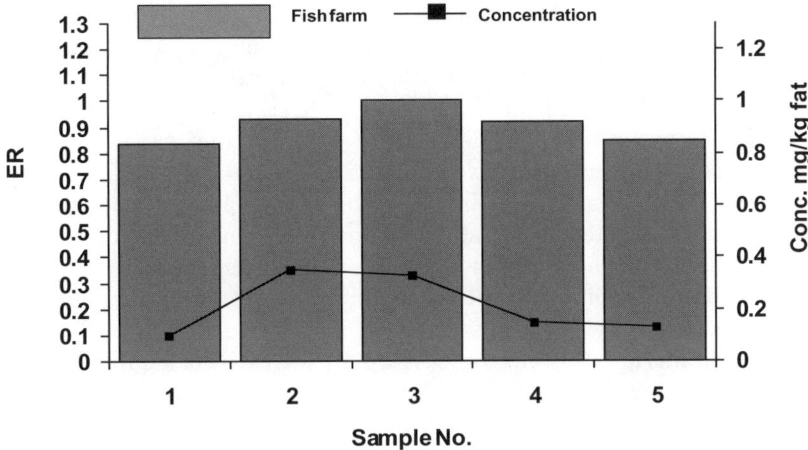

Fig. 3.16. Enantiomeric ratios [ER; (–)-/(+)-enantiomer] and concentrations of bromocylen in fish samples from Danish fish farms; from [50]

berger et al., gave rise to the enantioselective degradation of bromocyclen. As for the samples from a fish farm, a correlation between the enantiomeric ratios and the concentration can be inferred (Fig. 3.16). However, caution has to be applied when interpreting this result from a statistical point of view. The observed correlation is based on a consistent but small data set, and it is, therefore, not statistically significant. However, the observed enantiomeric ratios in the fish samples imply that at least the (–)-enantiomer can be metabolised to some degree.

The source of contamination which led to the high concentration in this matrix was unclear. It could not be ruled out that effluents of municipal sewage treatment plants into the River Stör could have been one source of contamination for the three fish caught in this river. However, the severe impact of bromocyclen on fish stemming from fish farms remains unclear thus far. In order to gain deepened insight into potential sources, Bethan et al. [49] carried out a comprehensive study which included the analysis of bromocyclen in water and fish samples from different stations of the River Stör. Moreover, samples from the influent as well as from the effluent of various sewage plants along the river were collected. Particular emphasis was placed on the question as to how bromocyclen gets into the River Stör and if the enantiomeric ratios determined in the fish samples correlate with the concentrations. The latter aspect was intended to verify the hypothesis of Pfaffenberger et al. outlined above.

The water samples (already discussed in Sect. 3.1.2) were collected between May and August in 1995 in the River Stör. The fish samples, i.e., two brook trout (*Salmo trutta fario* L.) and eight bream (*Abramis brama orientalis* BERG), were also caught in the River Stör between July and August 1995. In addition, three samples of the effluent from sewage plants in the cities of Hamburg, Itzehoe and Neumünster were taken during this period. Furthermore, two samples of the effluent as well as the influent from the sewage treatment plants in Neumünster and Itzehoe were collected in the autumn of 1996. The sample preparation and the clean-up procedure are described elsewhere [49, 66].

The concentrations in the ten fish samples varied between 0.01 and 0.24 mg/kg fat and showed the same general characteristics in the decreasing trend towards the estuary as those received from the water samples (see Sect. 3.1.2). High concentrations were found in the trout that had been caught in Neumünster downstream of the influx of the sewage plant's wastewater, while the fish samples collected at the estuary of the river showed low concentrations. However, it has to be noted that the bromocyclen concentration in fish reflects a long-term accumulation, while the water samples represent the state during the sample collection.

The enantiomeric ratios of bromocyclen for the extracts from two trout and eight bream samples from the River Stör are summarised in Table 3.12. In the muscle tissue of the bream, the ER values turned out to be higher than 1 and varied between 1.08 and 2.13. The values imply that the (+)-enantiomer was preferentially degraded, i.e., at least the (+)-enantiomer can be metabolised, or, alternatively, the (−)-enantiomer was preferentially accumulated. This result appears to be at variance with the results from a previous investigation of rainbow trout from a Danish fish farm that led to the contrary assumption (see above; [50]). However, the two trout included in the study by Bethan et al. also showed enantiomeric ratios smaller than one. Therefore, it can be tentatively assumed that bream and trout show different metabolic pathways with regard to bromocyclen contamination. Moreover, a correlation between the enantiomeric ratios and the concentrations can be inferred from the data set of Bethan et al. (Fig. 3.17). Though their result is based on two trout and eight bream samples only, it is evident that higher concentrations correlate with a higher relative concentration of the (+)-enantiomer. It is worth noting that to date nothing is known about the toxicity of the single enantiomers of bromocyclen, which would be useful for a further interpretation of these results.

Toxaphene is a broad-range pesticide, which was used primarily on cotton and soy bean in the south-eastern United States. Its use was banned by the US Environmental Protection Agency in 1986 because of its persistence. However, residues of technical toxaphene are still found in biota far removed from sites of its former application. Technical toxaphene is synthesised by the exhaustive chlorination of camphene, followed by a Wagner-Meerwein rearrangement under the usual experimental conditions giving rise to the formation of polychlorinated 1,7,7-trimethylbicyclo[2,2,1]heptanes (bornanes). Theoretically, 32767 chlorinated bornanes are conceivable, 16128 of which exist as enantiomeric pairs, while only 511 are achiral [74, 177]. As a consequence, toxaphene is an extremely complex mixture of hexa- to decachlorinated bornanes, bornenes, camphenes and camphedienes sold under various brand-names (Storbane, Melipax). In technical mixtures, much less congeners, some hundred derivatives have been detected thus far. The nomenclature of this class of compounds is unsatisfactory, because the IUPAC names are very long and can hardly be used in a manuscript. Therefore, several suggestions have been made for abbreviations

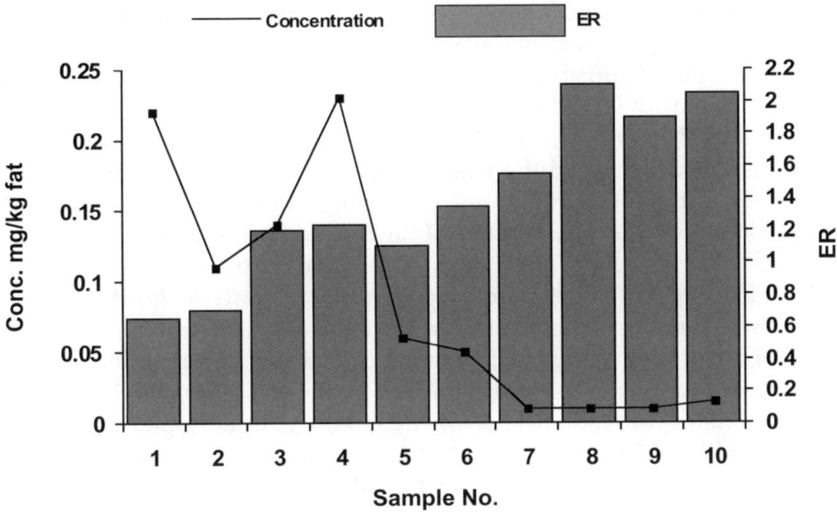

Fig. 3.17. Concentrations and enantiomeric ratios of bromocyclen in two trout (F1, F2) and eight bream sample (F3 to F10) extracts; from [49]

that are more applicable, two of which seem to dominate, i.e., the Parlar numbers [74] and the systematic code names proposed by Andrews and Vetter [75]. These will also be used in the present monograph. In addition, suggestions by Oehme and Kallenborn [76] and by Nikiforov et al. [77] are worth mentioning. As few of its numerous congeners have been synthesised thus far, toxaphene can only be analysed effectively with a selective analytical technique. In addition to the large number of congeners, the following problems have to be overcome: Firstly, gas chromatographic enantiomer separation of toxaphenes requires relatively high temperatures, which turned out to be unfavourable to a good chromatographic performance [74]. Secondly, Buser and Müller [71] presented experimental evidence that nonracemic congeners may be present in technical mixtures of toxaphenes. At first, these reports caused some scepticism among scientists. However, recently, Vetter et al. [27] were able to isolate the heptachlorobornane B7–1453 from the technical product Melipax (produced in the former German Democratic Republic), which showed different peak abundance of the enantiomers (ER 1.26±0.03). Furthermore, the authors excluded interferences and coelution with other compounds. Therefore, they conclude that B7–1453 and possibly other chlorobornanes are present in a nonracemic composition in technical mixtures. An explanation is provided by the use of natural precursors (α-pinene, terpene and camphene) in the toxaphene synthesis, which may give rise to chiral induction in the course of the preparative pathway. As a consequence, caution has to be applied when interpreting enantiomeric excesses of toxaphene congeners determined in biota, because deviations from a racemic composition in biological samples may not only arise from enantioselective transformation processes.

The first to report enantiomeric excesses of two major abundant chlorobornanes in biological samples were Kallenborn et al. [70] and Buser and Müller [51]. In both studies the same sample was used, i.e., blubber of a harbour seal (*Phoca vitulina* L.), from which the major chlorobornanes had been isolated. B9–1679 (Parlar50) showed good agreement with ERs of 1.06 vs. 1.08 (Table 3.13), but B8–1413 (Parlar26) deviated a little with an ER of 1.02 vs. 1.12. In the latter case, baseline separation of the enantiomers was not achieved in both studies, which might explain the larger discrepancy of the results. Furthermore, Buser et al. investigated Baltic herring (*Clupea harengus* L.) and harp seal (*Phoca groenlandica* ERX.) from Greenland, and Adelie penguin (*Pygoscelis adeliae* L.) from Antarctica. These samples were prepared at the Institute of Environmental Chemistry, University of Umeå, Sweden. It should be noted that the enantiomeric ratios for these biota sample extracts summarised in Table 3.13 have to be compared with the value for the technical toxaphene mixture. For example, for the congener B9–1679 (Parlar50), an ER of 1.13±0.02 was reported, i.e., a value that differs slightly from a racemic mixture. Therefore, the ERs determined for the warm-blooded animals are less significantly larger than the value in the technical mixture, as usually expected from technical compounds. It is worth noting that the congener B9–1679 (Parlar50) is a major component in all aquatic biota samples, but only a minor component of the technical product.

A very comprehensive study was carried out by Alder et al. [72], who analysed eight fish samples including six different species (Table 3.13). The enantiomeric ratios ranged between 0.91 and 1.13, i.e., when taking into account the experimental error of the method, only small differences from racemic distributions were encountered, and no species-dependent transformation can be inferred from the data set.

Vetter et al. [27, 73] analysed three Weddell seals (*Leptonychotes weddellii* LES.) and one Leopard seal (*Hydrurga leptonyx* BLA.) with regard to enantiomeric ratios of B8–1413 (Parlar26), B9–1679 (Parlar50), and B8–2229 (Parlar44). For all animals the second-eluting enantiomer of B8–2229 was below 40% of the first-eluting one (Table 3.13). Vetter et al. also isolated B7–1453 from a Baltic cod (*Gadus morhua* L.) liver extract and determined an ER-value of 1.0. This value has to be compared with the isolate from the technical Melipax mixture, which exhibited an ER value of 1.26±0.03, as mentioned above. In this special case, i.e., application of an enantio-enriched toxaphene congener, the alteration of the ER was due to a faster transformation of the first-eluting enantiomer, which finally led to an ER of 1.0. This spectacular result, enantioselective transformation leading to equal amounts of both enantiomers, is the result of biotransformation of enantiomers with different reaction speeds which accidentally led to an ER of 1.0. The interpretation of this result would be erroneous if synthetic (racemic) standards had been used as a reference instead of the B7–1453 isolate from Melipax. It cannot be excluded that a similar situation may be encountered in connection with other toxaphene congeners. Therefore, caution has to be applied, when interpreting enantiomeric ratios of toxaphene congeners determined from environmental samples. In a subsequent investigation, Vetter and co-work-

Table 3.13. Systematic code names, Parlar numbers, and enantiomeric ratios of toxaphenes determined in extracts of biota tissues (f=female, m=male, ad=adult)

Biota/tissue	Enantiomeric ratios							Ref(s)
	B8–1413 Parlar 26	B9–1679 Parlar 50	B8–2229 Parlar 44	B8–789	B9–1025	B7–1453	B8–1412	
Hake liver	0.87	0.89		0.14	0.35			[165]
Dolphin blubber (40-y-old male)	0.96	0.94		<0.05	0.30			
Dolphin blubber (10-y-old male)				<0.05	0.25			
Herring (Baltic Sea)	0.95±0.03	1.08±0.03						[72]
Saithe (N. Atlantic)	1.08±0.03	1.07±0.03						
Farmed salmon (Norway)	1.02±0.06	1.13±0.04						
Redfish (N. Atlantic)	1.1±0.1	1.08±0.07						
Herring (North Sea)	1.06	1.08±0.07						
Mackerel (N. Atlantic)	0.98±0.03	1.08±0.03						
Mackerel (North Sea)	1.13±0.12	1.11±0.03						
Halibut (N. Atlantic)	0.91±0.08	1.08±0.3						
Monkey (adipose)	1.29±0.08	1.08±0.03						[72]
Human milk A	1.28±0.09	1.06±0.03						[72]
Human milk B	1,28±0.04	1.17±0.03						
Human Milk C	1.17±0.1	1.23±0.06						
Human milk D	1.23±0.1	1.21±0.04						
Human milk, pooled sample (n=9)	1.07±0.05	1.33±0.03						
Harbour seal	1.02	1.06	–			–	–	[70]
Tech toxaphene mixt		1.13±0.02	–			–	–	[51]
Harbour seal	1.12±0.04	1.08±0.04	–			–	–	[51]
Baltic herring		0.98±0.08	–			–	–	[51]
Harp seal		1.19±0.03	–			–	–	[51]
Adelie penguin		1.34±0.02	–			–	–	[51, 71]
Weddell Seal, f, subadult	1.1	1.2	3.4			–	–	[27, 73]
Weddell Seal, m, ad.	1.2	1.3	2.6			–	–	[27, 73]
Weddell Seal, m, ad.	1.1	1.3	2.8			–	–	[27, 73]
Leopard seal, unknown, ad.	1.3	1.6	4.3			–	–	[27, 73]
Baltic cod liver						1.0	–	[27, 73]
Tech toxaphene mixt						1.26±0.03	–	[27, 73]

Table 3.13. (continued)

Biota/tissue	Enantiomeric ratios							Ref(s)
	B8–1413 Parlar 26	B9–1679 Parlar 50	B8–2229 Parlar 44	B8–789	B9–1025	B7–1453	B8–1412	
Baltic cod liver							0.5±0.15	[98]
Grey seal							0.4	[98]
Weddell seal							0.4	[98]
Techn. mixture TC6	1.04±0.03							[51]
Antarct penguin TC6	0.91±0.02							[51]
Techn. mixture TC3	1.00±0.03							[51]
Baltic salmon TC3	0.3							[51]
Techn. mixture TC9	1.00±0.03							[51]
Baltic salmon TC9	1.4							[51]
Techn. mixture TC1	1.02±0.02							[51]
Antarct penguin TC1	0.74±0.03							[51]

ers confirmed the enantiomeric ratios published for the four toxaphene congeners (B7–1453 [74], B8–1413 [27, 73], B8–2229 [73], and B9–1679 [27, 73]) by carefully studying and excluding coelution effects and other artifacts that may have been caused, for example, in the case of cGC/ECNI-MS application [99].

Recently, Klobes et al. [98] succeeded in separating the enantiomers of toxaphene congener B8–1412, a major component in biota, using a 30 m long column coated with 35% heptakis(6-*O-tert*-butyldimethylsilyl-2,3-di-*O*-methyl)-β-cyclodextrin diluted in OV-1701. In all the cod liver samples from different areas of the Baltic Sea, enantiomeric ratios of B8–1412 were significantly less than 1 (Table 3.13). An ER of 0.4 was found both in blubber extracts from a grey seal (*Halichoerus grypus* FABR.) from Iceland and in blubber extracts from a Weddell seal (*Leptonychotes Weddeli* LES.). Although the number of samples investigated thus far is low, the results obtained by Klobes et al. clearly indicate an enantio-enriched B8–1412 in all samples.

Garrison et al. were not able to verify any enantioselectivity for the transformation of *o,p'*-DDT to *o,p'*-DDD using the *Elodea*/water reaction medium described in Section 3.1.1 [68]. By way of contrast, enantioselectivity was proved by these authors for the occurrence of *o,p'*-DDD in fish tissue [68]. Extracts from 21 fish captured from a river where the sediment has a history of severe DDT contamination from a long-defunct pesticide manufacturing plant were analysed by enantioselective cGC and CE, respectively, for *o,p'*-DDD. The fish species were channel catfish (*Ictalurus punctatus* RAF.), buffalo (*Ictiobus cyprinellus* VAL.) and largemouth bass (*Micropterus salmoides salmoides* N&M).

As Garrison et al. used a modified γ-cyclodextrin phase as chiral selector, they assumed that the (–)-enantiomer eluted first, in line with suggestions made by Buser and Müller [175], who separated these enantiomers with a γ-cyclodextrin-based HPLC column. Among the 21 samples, only two showed positive ER values (both values were 1.10). The range of ratios of the 19 samples with a negative ER was 0.25 to 0.98, but 14 of these fell between 0.40 and 0.79. Apparently, the biological uptake mechanism or some membrane transport process, or both, favoured the (−)-enantiomer, or else the (+)-enantiomer was metabolised faster by the fish. However, there was no correlation between fish species and direction or degree of enantioselectivity. Basically, it cannot be excluded that the o,p'-DDD was partially degraded by an enantioselective process in the environment, probably in the sediment. Garrison et al. intended to verify this hypothesis by additional analyses of o,p'-DDD enantiomers in the sediment samples that were also collected.

Polychlorinated biphenyls (PCBs), highly persistent lipophilic industrial chemical compounds, have been in focus as a major and ubiquitous environmental contaminant for more than three decades. Basically, 209 structurally and chemically related congeners are conceivable, representing a wide range of physicochemical properties. PCBs have been used as industrial fluids, flame retardants, diluents, hydraulic fluids and dielectric fluids for capacitors and transformers. Extensive contamination with PCBs has occurred during the period of their industrial use, from the early 1930s until the 1980s. Although PCBs have been banned from industrial application since the early 1980s, they are still entering the environment. Primary sources are leakages from old so-called *closed systems*, such as capacitors and transformers, and the disposal of materials contaminated with PCBs, such as old paints, painted construction materials, lubricant oils, sealing material and fire retardants in old fire extinguishers. Furthermore, a number of secondary sources of PCBs can be identified including resuspended river sediments, leakages from dump sites, dumping of sewage sludges and long-range atmospheric transport.

Jensen and co-workers were the first to report PCBs in environmental extracts in the late 1960s [78], and subsequent analytical studies have demonstrated the presence of PCBs in almost every compartment of the global ecosystem including the air, water, sediments, fish, wildlife and humans. In particular, in the marine environment PCBs were not only encountered in seawater, but also in species of different trophic levels of the marine ecosystem [79–83].

Analytical methods for the determination of PCBs have improved in recent years, thus permitting the determination of most of the 150 congeners of technical PCB mixtures in environmental samples. In the last few years, increasing attention has been paid to the analysis of coplanar and *atropisomeric* congeners (see Sect. 1.2.2). The toxicological implications related to the structures of these two groups will be summarised in Section 4.4. Herein, emphasis will be placed on the analysis of atropisomeric PCBs in environmental samples.

Basically, 78 out of 209 PCB congeners display axial chirality in their nonplanar conformations, the so-called *atropisomerism*. In a pioneering account, Kaiser predicted that 19 PCBs, nearly all of which are present in commercial PCB formulations, exist as stable atropisomers at ambient temperatures due to restricted rotation around the central C–C bond of the biphenyl system ($G^* = 104$–242 kJ/mol; 25–58 kcal/mol) with "a bearing on the toxicity and metabolic interactions of these chemicals" [84, 171].

The first cGC separations of atropisomeric PCBs were published in 1993 and 1994 [20, 29, 31, 53], and in 1994 and 1995 the first reports on the enantioselective determination of an atropisomeric PCB in biological samples appeared [53, 83, 85]. Meanwhile, several enantiopure or enantio-enriched atropisomeric PCBs have been isolated by application of enantioselective HPLC and can thus be used as standard compounds [86–89], e.g., for the determination of the elution order of PCB atropisomers in enantioselective cGC [34].

With regard to enantioselective analysis of atropisomeric PCBs in marine and limnic biota tissues, only a few studies have been reported in the literature. In 1994, Hühnerfuss et al. [83, 85] reported for the first time about the identification of five atropisomeric PCBs in blue mussels (*Mytilus edulis* L.) on an achiral CP-Sil 5/C 18 CB fused-silica capillary column (Chrompack; length 100 m). Higher levels of PCB88, PCB149, PCB171, PCB174 and PCB183 were determined in spring as compared to autumn at six sampling sites in the Weser, Jade, and Elbe river estuaries (German Bight). Furthermore, the enantiomers of PCB149 in all mussels collected during the spring and autumn period were separated. The enantiomeric ratios ranged between 1.0 and 1.2, which implies a weak enantio-enrichment of the first-eluting enantiomer.

Blanch et al. [90] also used enantioselective cGC for the determination of three atropisomeric PCBs in liver samples of shark (*Centroscymnus coelolepis*, B.&C.). While PCB95 and PCB149 were present in racemic compositions, the second-eluting enantiomer of PCB132, i.e., (+)-PCB132, predominated in most of the samples (ER 0.75–0.89). Ramos et al. analysed nine atropisomeric PCBs in two otter samples [91]. The ERs were in part extremely high, although the report is lacking a detailed experimental description. Enantio-enrichment of PCB149 in blubber of an adult female harbour seal (*Phoca vitulina* L.) sample from Iceland was determined by Vetter et al. [32]. They report that the first-eluted peak was significantly higher than the second one. In a later study, enantio-enriched PCB149 was also found in the blubber of further harbour seals as well as grey seals (*Halichoerus grypus* FABR.), and a caspian seal (*Phoca caspica* L.) [74].

Reich et al. [100] determined enantiomeric ratios of atropisomeric PCBs in dolphins (*Stenella coeruleoalba* ME.) from the Mediterranean Sea. The six dolphins were found dead along the Italian coast (Lygurian and Tirrhenian Seas) in the period 1989–1990. The enantiomeric ratios (defined as the relation of the first-eluting enantiomer to the second one) of the nine atropisomeric PCBs (84, 91, 95, 132, 135, 136, 149, 174, 176) obtained in the samples studied revealed that PCB95 (ER 0.71–1.07), PCB136 (ER 0.84–1.07), PCB174 (ER 0.58–1.16), and PCB176 (ER 0.72–0.97) were racemic or nearly racemic in almost all samples investigated by Reich et al. PCB95 exhibited an enantiomeric excess (*ee*) of 17% of the second-eluting atropisomer in two liver samples, and PCB174 showed an *ee* of 26.6% in one liver sample. PCB132 (ER 0.55–0.92), PCB135 (ER 0.63–0.76) and PCB149 (ER 0.58–0.91) revealed an *ee* of the second-eluting enantiomer in almost all sample extracts. No *ee* was found for PCB132 and PCB149 in one sample each.

The differences observed in the enantiomeric ratios of the atropisomeric PCBs could not be explained by the relationship between structure and metabolism. PCB95, PCB132, PCB135, PCB149, PCB174, and PCB176 belong to the readily metabolisable PCBs. They possess vicinal hydrogen atoms in both *ortho/meta-* and *meta/para*-positions (PCB132), in two *meta/para*-positions (PCB95, PCB136) or in one *meta/para*-position (PCB135, PCB149, PCB174, PCB176). It is, therefore, not possible, on the basis of its structure, to explain why PCB95 (with two vicinal H atoms in *meta/para*-positions) only shows slight enantiomeric enrichment, while PCB149 (with only one free *meta/para*-position) exhibits higher enantiomeric enrichment. Thus, Reich et al. conclude that the differences found in the metabolic degradation pathway between the two atropisomers of these PCBs could be better explained by the enantioselective character of the enzymatic biodegradation process as suggested by Faller et al. [8].

3.2.1.2
Terrestrial Ecosystem

Enantioselective analyses of biota from terrestrial ecosystems have so far largely focused on the questions: can the conclusions drawn from marine biota analyses with regard to enantioselective transformation of xenobiotics be transferred to terrestrial animals? Can species-dependent and/or concentration-dependent effects be observed? The first to address these questions were Möller et al. [44] who determined the enrichment of α-HCH enantiomers as determined in fat, liver, and brain tissue samples of sheep (*Ovis ammon* L.) bred in the northern German state, Schleswig-Holstein (Table 3.14; Fig. 3.18). In the fat and liver a depletion of the (+)-enantiomer was observed, while in the brain vice versa the (+)-enantiomer was dominant. The latter aspect will be pursued in Section 3.2.2 (enantioselective permeation through the blood-brain barrier). These values were compared with those from marine biota like blue mussels (*Mytilus edulis* L.; ER between 0.67 and 0.89), flounder (*Platychthys flesus* L.; ER 0.80–0.94), common Eider duck (*Somateria mollissima* L.; liver: ER 1.4–∞; kidney: ER 1.6; muscle: ER 7.0), and harbour seals (*Phoca vitulina*; blubber: ER 1.2–4.5; brain ER 7.9–∞). Möller et al. concluded that the enrichment of the α-HCH enantio-

Table 3.14. Comparison of the residue contents and enantiomeric ratios (ER) of α-HCH in sheep liver, fat and brain samples from Schleswig-Holstein; from [44, 95]

Sample No.	Liver		Fat		Brain	
	Conc. (µg/g EOM)	ER (+)/(−)	Conc. (µg/g EOM)	ER (+)/(−)	Conc. (µg/g EOM)	ER (+)/(−)
1	0.010	0.96	0.008	0.85	<0.001	3.76
2	0.006	0.69	0.004	0.64	<0.001	3.00
3	0.007	0.89	0.005	0.70	<0.001	2.13
4	0.011	0.75	0.013	0.56	<0.001	1.86
5	0.015	0.85	0.012	0.96	<0.001	1.78
6	0.013	0.75	0.013	0.64	<0.001	1.50
7	0.012	0.79	0.009	0.82	<0.001	1.38

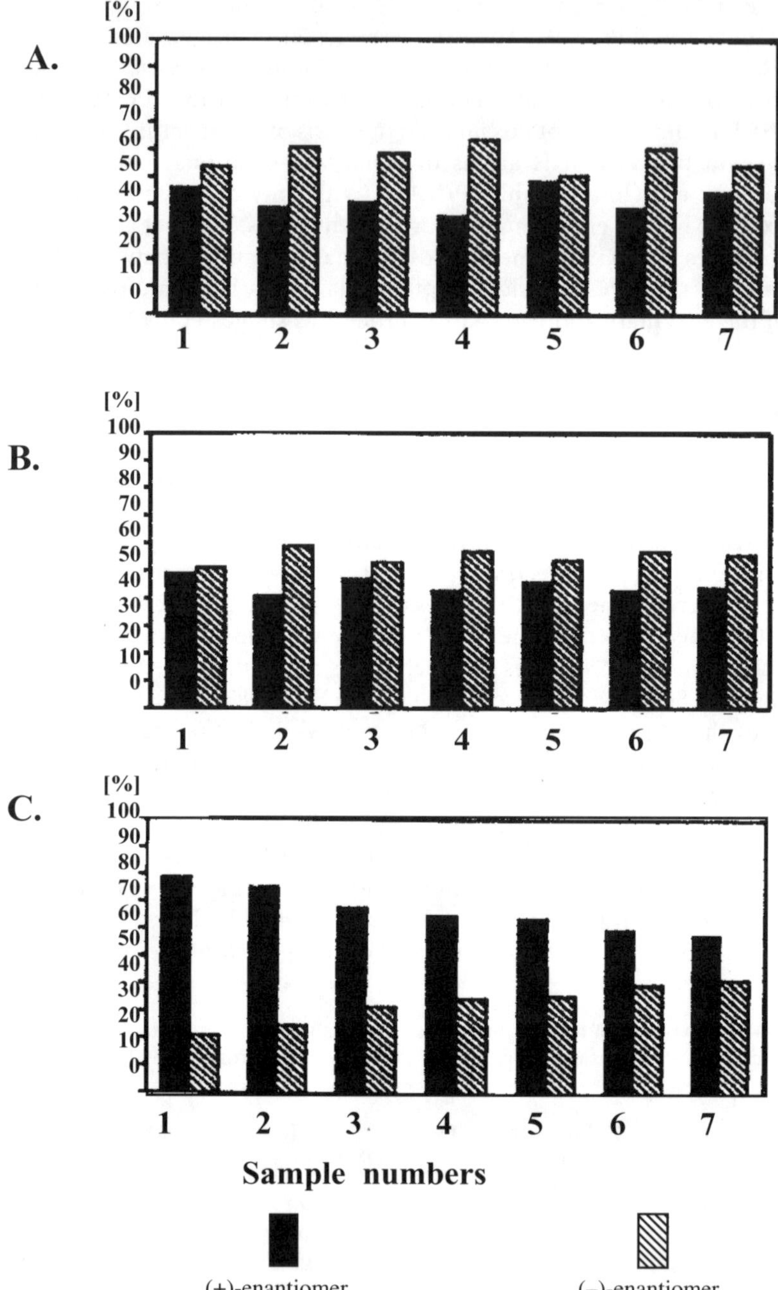

Fig. 3.18A–C. Enrichment of α-HCH enantiomers (%) as determined in **A** fat, **B** liver, and **C** brain tissue samples of sheep bred in the northern German state Schleswig-Holstein; from [44]

mers in sheep fat and liver tissues is different from those found in the respective tissues of marine biota of high trophic levels, and they assumed that the enzymatic transformation pathways in the livers of the terrestrial (sheep) and marine animals (harbour seals) of higher trophic levels are different. However, it is worth noting that the occurrence of β-pentachlorocyclohexene in all tissues compared in the study by Möller et al. indicated the usual transformation mechanism, i.e., *trans*-dehydrohalogenation of α-HCH.

The impact of harmful substances on game animals from various regions of the northern German state Schleswig-Holstein and the southern German state Baden-Württemberg was investigated in the course of a long-term program. The studies included different species of wild animals. Special emphasis was placed on the residue contents of organochlorine compounds in the muscle and liver tissues of game animals. The residue contents of α-HCH in liver tissues for both regions showed a surprising result. The roe-deer (*Capreolus capreolus* L.) livers exhibited remarkably high α-HCH contents. Furthermore, high concentrations of chlordane isomers were detected in the liver samples. This is insofar noteworthy, as the application of these compounds was banned in Germany as early as 1974.

In the study by Pfaffenberger et al. [43, 96], liver samples of roe-deer were analysed for their content of α-HCH, heptachlor *exo*-epoxide, and oxychlordane. Particular emphasis was placed on the question as to whether or not a correlation between the concentrations of these chiral compounds and their enantiomeric ratios can be inferred, independent of the respective environmental milieu. In order to answer this question, samples from two different German regions, Schleswig-Holstein and Baden-Württemberg, were investigated, where the concentrations of several contaminants, e.g., α-HCH and chlordane isomers, in liver tissue of roe-deer are of the same order of magnitude. The roe-deer (*Capreolus capreolus* L.) from northern Germany were shot during the hunting seasons 1992 and 1993 in the southern part of Schleswig-Holstein. The roe-deer from the southern German state Baden-Württemberg were shot during the hunting seasons 1989 and 1990. The liver samples were stored in a refrigerator at about 248 K prior to sample preparation (for details see [43]).

Eight liver samples of roe-deer from Schleswig-Holstein and nine liver samples of roe-deer from Baden-Württemberg were analysed. In all eight liver samples from Schleswig-Holstein, remarkably high concentrations of α-HCH, between 20 and 140 µg/kg fat, were determined (Table 3.15). This phenomenon was not observed for kindred animal species, e.g., red-deer and fallow-deer. A direct uptake of technical α-HCH is unlikely, because this would also result in increased β-HCH concentrations, which was not the case. One reason for the high α-HCH concentrations could be an isomerisation of γ-HCH to α-HCH in the liver. But this is hypothetical and could not be derived from the data set obtained by Pfaffenberger et al. The α-HCH concentrations in the liver samples from Baden-Württemberg were within the same order of magnitude as the samples from Schleswig-Holstein. The concentrations varied between 56 and 300 µg/kg fat (Table 3.15), i.e., no noteworthy regional differences existed between

Table 3.15. Comparison of the residue contents and enantiomeric ratios (ER) of α-HCH in roe-deer liver samples for the regions Schleswig-Holstein and Baden-Württemberg; from [43]

	Schleswig-Holstein			Baden-Württemberg	
Sample No.	Conc. (µg/kg EOM)	ER (+)/(−)	Sample No.	Conc. (µg/kg EOM)	ER (+)/(−)
1	60	0.15	9	56	0.23
2	140	0.06	10	63	0.35
3	80	0.06	11	82	0.23
4	100	0.03	12	150	0.16
5	100	0.04	13	220	0.17
6	50	0.07	14	178	0.12
7	40	0.40	15	119	0.18
8	20	0.35	16	267	0.09
			17	300	0.13

Schleswig-Holstein and Baden-Württemberg concerning the α-HCH contents in the liver samples.

For all liver samples the enantiomeric ratios of α-HCH were determined. In all instances, the ratios (+)-α-HCH/(−)-α-HCH turned out to be significantly lower than one. The values varied between 0.03 and 0.40 for the samples from Schleswig-Holstein, and between 0.09 and 0.35 for the samples from Baden-Württemberg (Table 3.15). These values imply that the (+)-enantiomer was preferentially degraded. Furthermore, the results show that the enantiomeric ratios of α-HCH in roe-deer liver tissues are inverse to those found in the respective tissues of marine biota of higher trophic levels. On the other hand, in liver tissues of sheep, i.e., a terrestrial species of nearly the same trophic level as roe-deer, preferential degradation of the (+)-enantiomer or vice versa enrichment of the (−)-enantiomer was also observed, as discussed above (Fig. 3.18). Therefore, it can be tentatively assumed that different enzymatic systems in the livers of terrestrial and marine animals of higher trophic levels give rise to different enantiomeric enrichments.

For a more detailed interpretation, Pfaffenberger et al. [43, 96] applied the Spearman rank correlation test to the data set [93], with the aim of checking whether or not a correlation, described by the Spearman rank correlation coefficient r_s, exists between the concentrations and the enantiomeric ratios. For both data sets, confidence limits of 95% were assumed. For the data from Schleswig-Holstein ($r_s = -0.35$), a weak negative, but statistically not significant, correlation was obtained, whereas for the data from Baden-Württemberg ($r_s = -0.72$) a strong negative and statistically significant correlation was observed (Fig. 3.19). Moreover, it should be noted that for a further comprehensive statistical interpretation the data set was not large enough. However, a first indication can be inferred from the results that higher concentrations of α-HCH may result in a stronger decomposition of the (+)-enantiomer. This holds at least for the samples from Baden-Württemberg.

A

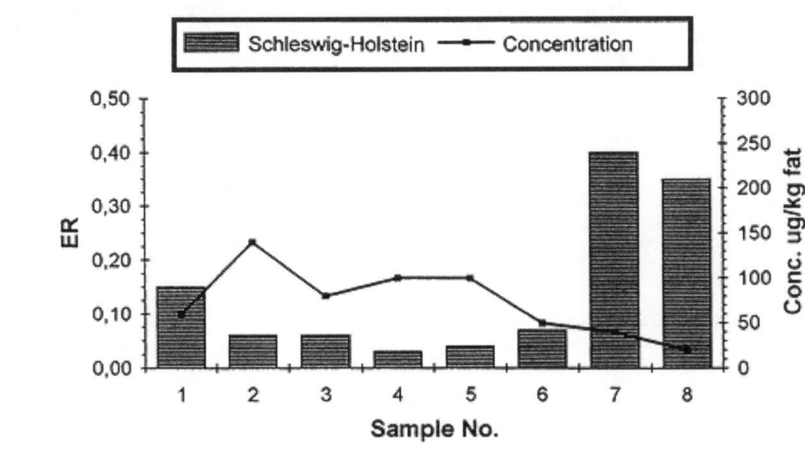

B

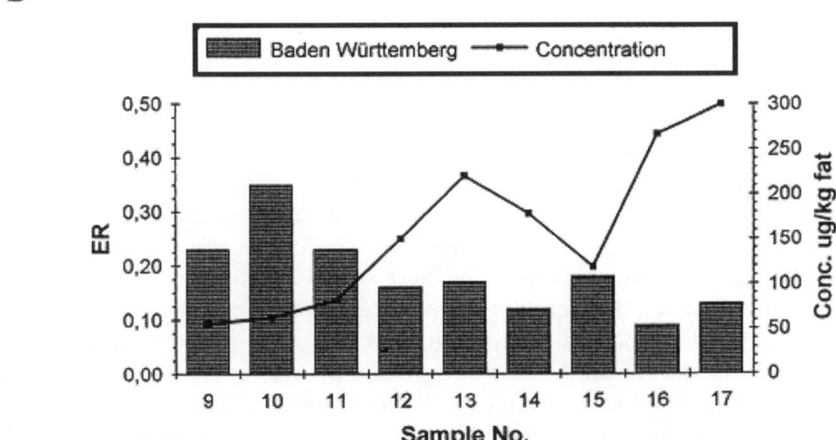

Fig. 3.19A, B. Enantiomeric ratios (ER; (+)-/(–)-enantiomer) and concentrations (µg/kg fat) of α-HCH for **A** roe-deer liver samples from Schleswig-Holstein (Spearman rank correlation coefficient $r_s = -0.35$) and **B** roe-deer liver samples from Baden-Württemberg ($r_s = -0.72$); from [43]

In the samples from Schleswig-Holstein and Baden-Württemberg, noteworthy concentrations of oxychlordane and heptachlor *exo*-epoxide were also found. In line with previous reports from other authors for terrestrial animals of higher trophic level [94], from both possible isomers of heptachlorepoxide, the *exo*- and *endo*-isomer, only the heptachlor *exo*-epoxide was found in the liver of roe-deer. In all eight liver samples from Schleswig-Holstein, unusually high concentrations of oxychlordane and of heptachlor *exo*-epoxide were determined. The concentrations vary between 10 and 60 µg/kg fat for oxychlordane and between 10 and 100 µg/kg fat for heptachlor *exo*-epoxide (Table 3.16). This is re-

Table 3.16. Concentrations (µg/kg fat) and enantiomeric ratios (ER) of oxychlordane and heptachlor *exo*-epoxide in roe-deer livers from Schleswig-Holstein and Baden-Württemberg; from [43]

Sample No.	Oxychlordane		Heptachlor *exo*-epoxide	
	Conc.(µg/kg fat)	ER (+)/(−)	Conc. (µg/kg fat)	ER (+)/(−)
Schleswig-Holstein				
1	40	9	10	1
2	60	12	40	2
3	50	14	100	6
4	30	11	40	2
5	60	17	90	7
6	40	12	70	9
7	30	11	30	5
8	10	7	20	5
Baden-Württemberg				
9	16	−	17	−
10	53	−	21	−
11	60	−	81	−
12	82	27	45	2
13	46	19	140	3
14	31	−	61	−
15	13	−	10	−
16	38	20	56	2
17	100	18	236	3

markable, because the application of these soil insecticides was stopped in the mid-1970s. However, this result is a consequence of the negative environmental properties of such chlorinated compounds, i.e., slow degradation and high bio-accumulation. The concentrations of both compounds in the liver samples from Baden-Württemberg were, as already discussed for α-HCH, in the same order of magnitude. The concentrations vary between 13 and 100 µg/kg fat for oxychlordane, and between 10 and 236 µg/kg fat for heptachlor *exo*-epoxide (Table 3.16). These values are a further indication that there are no great differences between the regions Baden-Württemberg and Schleswig-Holstein with regard to the contamination levels of oxychlordane and heptachlor *exo*-epoxide.

In order to gain deeper insight into the enantioselective transformation of oxychlordane and heptachlor *exo*-epoxide, their enantiomeric ratios (ERs) were determined by enantioselective cGC. This was not possible for all samples from Baden-Württemberg (in Table 3.16 indicated with "-") due to coeluting matrix peaks, which could not be sufficiently well separated by the HPLC clean-up. The assignment of the order of elution of enantiomers is based on optical rotation measurements after preparative enantiomeric resolution by packed-column GC [25]. In all samples, with the exception of heptachlor *exo*-epoxide in sample No 1, the (−)-enantiomer was preferentially degraded. The same trend was observed for the (−)-enantiomer of oxychlordane, where the enantiomeric dis-

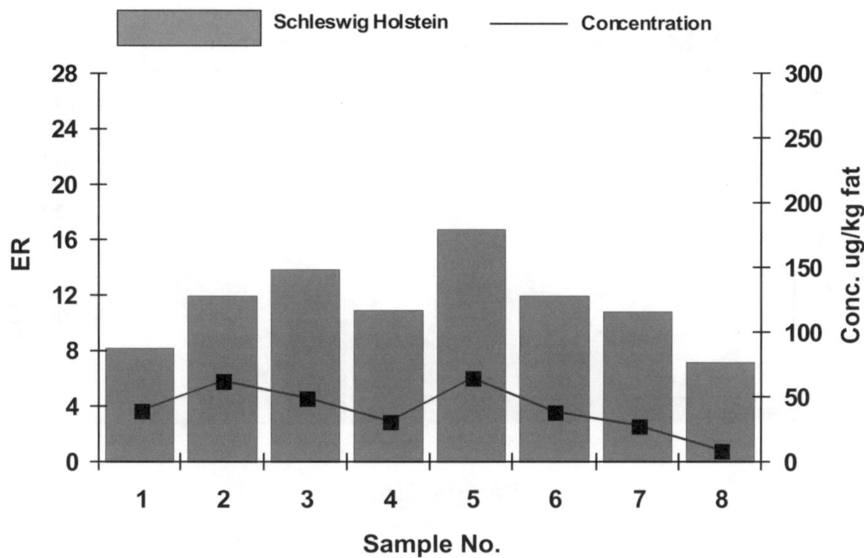

Fig. 3.20. Enantiomeric ratios [ER; (+)-/(–)-enantiomer] and concentrations (µg/kg fat) of oxychlordane for roe-deer liver samples from Schleswig-Holstein (Spearman rank correlation coefficient $r_s = 0.92$); from [43]

crimination appeared to be stronger in the liver samples from Baden-Württemberg than in the samples from the region Schleswig-Holstein. By way of contrast, the (–)-enantiomer of heptachlor *exo*-epoxide was degraded slightly more strongly in the liver samples from Schleswig-Holstein. As it cannot be expected that different enzymatic systems are encountered in the roe-deer livers of the two German regions, other explanations for this regional change have to be checked, e.g., different physical conditions of the animals, which in turn may be influenced by different diseases or by stress induced by environmental pollutants.

As described for α-HCH, the Spearman rank correlation test was applied to the data set from Schleswig-Holstein (top of Table 3.16) with the aim of checking whether or not a correlation exists between the concentrations and the enantiomeric ratios of the chlordane metabolites. The limitations are the same as mentioned above. The data set from Baden-Württemberg was not large enough for this correlation test. For oxychlordane, a very strongly positive significant correlation ($r_s = 0.92$; Fig. 3.20) and for heptachlor *exo*-epoxide a strongly positive significant correlation ($r_s = 0.76$; Fig. 3.21) were obtained. These values imply that higher concentration levels of oxychlordane and heptachlor *exo*-epoxide in the roe-deer livers may result in a faster decomposition of the (–)-enantiomer or in a preferential formation of the respective (+)-enantiomer. However, it should be noted that the observed correlation is based on a consistent but small data set. Therefore, caution has to be applied when generalising this result for other animals.

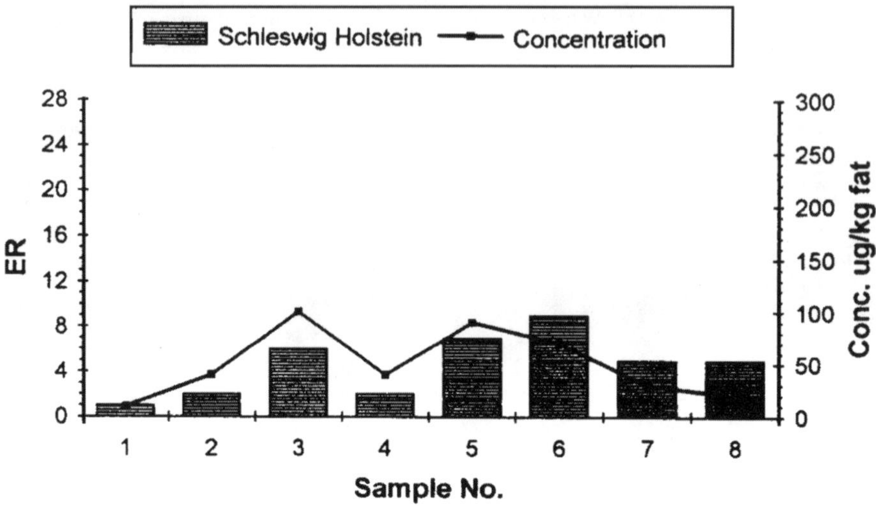

Fig. 3.21. Enantiomeric ratios ([ER; (+)-/(–)-enantiomer] and concentrations (µg/kg fat) of heptachlor *exo*-epoxide for roe-deer liver samples from Schleswig-Holstein (Spearman rank correlation coefficient $r_s = 0.76$); from [43]

Table 3.17. Comparison of the residue contents (µg/kg fat) and enantiomeric ratios (ER) of α-HCH, oxychlordane, and heptachlor *exo*-epoxide in hare (*Lepus europaeus* PAl.) liver samples from Schleswig-Holstein; from [25, 64, 96]

Sample No.	α-HCH		Oxychlordane		Heptachlor *exo*-epoxide	
	Conc. (µg/kg fat)	ER (+)/(–)	Conc. (µg/kg fat)	ER (+)/(–)	Conc. (µg/kg fat)	ER (+)/(–)
1	10	0.8	40	1.1	80	3.3
2	10	0.8	20	1.0	80	2.5
3	10	1.5	40	1.3	70	3.7
4	10	1.2	40	1.3	100	2.6
5	10	0.8	40	1.5	90	3.2

In addition to roe-deer, the long-term study carried out in the northern German state Schleswig-Holstein also included hare (*Lepus europaeus* PAL.) [25, 64, 94, 96]. The concentrations of α-HCH, oxychlordane, and heptachlor *exo*-epoxide in hare liver samples ranged between 10 and 100 µg/kg fat (Table 3.17), i.e., within the same order of magnitude as reported for the roe-deer liver samples discussed above. Furthermore, the enantiomeric ratios were determined in hare liver extracts using a 1:1 mixture of OV-1701/octakis(3-*O*-butyryl-2,6-di-*O*-*n*-pentyl)-γ-cyclodextrin (Lipodex E) for α-HCH, and heptakis(2-*O*-methyl-3,6-di-*O*-*n*-pentyl)-β-cyclodextrin for oxychlordane and heptachlor *exo*-epoxide as chiral selectors. In the case of α-HCH, ER values smaller as well as larger than one were found, a phenomenon which presently cannot be explained, because the data set was too small (Table 3.17). For oxychlordane, ER values between 1.0

and 1.5 were obtained indicating a preferential enzymatic transformation of the (–)-enantiomer. An even more pronounced transformation of the (–)-enantiomer can be inferred from the ER values for heptachlor *exo*-epoxide, ranging between 2.5 and 3.7. This is insofar remarkable as the results for roe-deer liver extracts indicated a stronger transformation of the (–)-enantiomer of oxychlordane as compared with that of heptachlor *exo*-epoxide. This phenomenon can be tentatively explained by different enzymatic systems prevalent in roe-deer and hare livers, respectively. Furthermore, it cannot be excluded that additional parameters, like different food habits, trophic levels, and the physiological conditions of these two species, may also supply a partial explanation.

The first to attempt the determination of enantiomeric excesses of PCB enantiomers in terrestrial ecosystems, i.e., in human milk, were Glausch et al. [53]. However, although they basically were able to separate the enantiomers of standard compounds of the chiral PCBs 95, 132 and 149 by multidimensional gas chromatography (MDGC) on Chirasil-Dex columns, none of the first trials indicated significant enantiomer enrichment in human milk. Subsequent further screening of PCB132 in human milk extracts suggested the enrichment of the second-eluted enantiomer. However, the unambiguous determination of PCB132 enantiomers turned out to be difficult. Reliable confirmation of the first results was attained by application of MDGC with two differently polar achiral stationary phases for pre-separation as well as cGC-MS in the SIM mode. As a result, enantiomeric ratios of PCB132 between 0.40 and 0.87 were observed in ten milk samples [54]. Since Haglund and Wiberg [34] used the same chiral selector as Glausch et al., the elution order of PCB132 published in their work allows the conclusion that, in the human milk samples investigated by the latter authors, (+)-PCB132 was the more abundant enantiomer. Furthermore, Ramos et al. analysed nine atropisomeric PCBs in a cow's milk sample [91]. The enantiomeric ratios in part were extremely high, but as no sufficient experimental details on this study were presented, it is difficult to make an assessment based on these results.

Approximately ten years after the discovery of PCBs in the environment [78], Jensen and Jansson reported on the identification of PCB methyl sulfones ($MeSO_2$-PCBs) in Baltic grey seal blubber [101]. About twenty years later, $MeSO_2$-PCBs were detected in fish, birds and mammals including humans (see refs in [35, 36]). More recently, some of the $MeSO_2$-PCBs in biota have also been observed to be selectively and strongly retained in the liver tissue of mammals including man. However, the mechanism for this selectivity is still unknown, although reversible protein binding plays a major role for their retention in the liver. The most important sulfones bound in mammalian liver are 3-$MeSO_2$-2,5,6,2',3',4'-hexachlorobiphenyl (abbreviated: 3–132; see [35]), 3-$MeSO_2$-2,5,6,2',4',5'-hexachlorobiphenyl (abbr. 3–149), and 4-$MeSO_2$-2,3,6,2',4',5'-hexachlorobiphenyl (abbr. 4–149). The $MeSO_2$-PCBs formed are persistent and only slightly less hydrophobic than their parent compounds which make them long-lasting contaminants in the biosphere. From a toxicological point of view, several of the 3-$MeSO_2$-PCBs have been shown by Kato and co-workers to induce strongly P-450 cytochrome enzymes such as P450 2B1, 2B2, 3A2 and 2C6 [102, 103]. As a consequence, it is tentatively assumed that a part of the toxic effects induced by PCBs in the environment may be subject to the presence of these PCB metabolites. Furthermore, the main metabolites mentioned above are chiral and, accordingly, enantioselective transformation as well as toxic impacts cannot be excluded. The latter aspect will be pursued in Section 4.4.

In order to gain a deeper insight into the enantioselective transformation of atropisomeric PCBs, the research groups of Bergman and Hühnerfuss [35, 36, 104] started a systematic joint study, which included:

- the separation of eight $MeSO_2$-PCB standards into their enantiomers using enantioselective cGC: 4–91, 4–95, 3–149, 4–149, 3–132, 4–132, 3–174 and 4–174 (for abbreviations, see [35]);
- the enantiomer separation of $MeSO_2$-PCBs in human liver sample extracts;
- the enantiomer separation of $MeSO_2$-PCBs in rat liver and adipose sample extracts;
- the enantiomer separation of $MeSO_2$-PCBs in rat lung sample extracts;
- the enantiomer separation of larger amounts of 3–132, 3–149 and 4–149 as well as their parent PCBs by enantioselective preparative HPLC; and
- an investigation of the enantioselective toxic effects of the $MeSO_2$-PCB enantiomers.

First results obtained from this ongoing joint endeavour were recently published [35, 36]: eight out of ten atropisomeric $MeSO_2$-PCBs (see above) previously found in environmental samples were separated into their enantiomers using a 10 m fused-silica capillary column coated with a 1:1 (w/w) mixture of OV-1701 and heptakis(6-*O-tert*-butyldimethylsilyl-2,3-di-*O*-methyl)-β-cyclodextrin (TBDMS-CD). The congener 3–95 was not included in the study because no standard compound was available. Furthermore, Ellerichmann and al. [35] reported the successful separation of the congener 3–91. However, in the course of subsequent investigations, Ellerichmann showed that another chiral component of the standard mixture, presumably 4-$MeSO_2$-2,2',3,3',6-penta-CB (i.e., 4–84), coeluted with the 3–91 peak thus giving rise to a misinterpretation of the original data set [172]. In general, the $MeSO_2$-PCBs exert strong interactions with the chiral cyclodextrin phase and, due to the rather low maximum temperature for the column (468 K), the retention times were long (from 50 min for 3–91 to 130 min for 4–174) despite the short length of the column. The peak width at half height was 30 s for the $MeSO_2$-pentaCBs and increased to up to 110 s for the $MeSO_2$-heptaCBs. Baseline separation was observed for most enantiomers and constitutional isomers (Fig. 3.22).

During the second part of the study, six human livers and two human lungs were obtained from the "Institut für Rechtsmedizin at the University of Hamburg" who performed autopsies on seven persons who had died due to heart failure or accidents [35]. The liver samples were analysed for content as well as for the enantiomeric excess of the eight $MeSO_2$-PCBs mentioned above. Two $MeSO_2$-PCBs (3–149 and 3–132) were detected in all liver samples by ECD, using

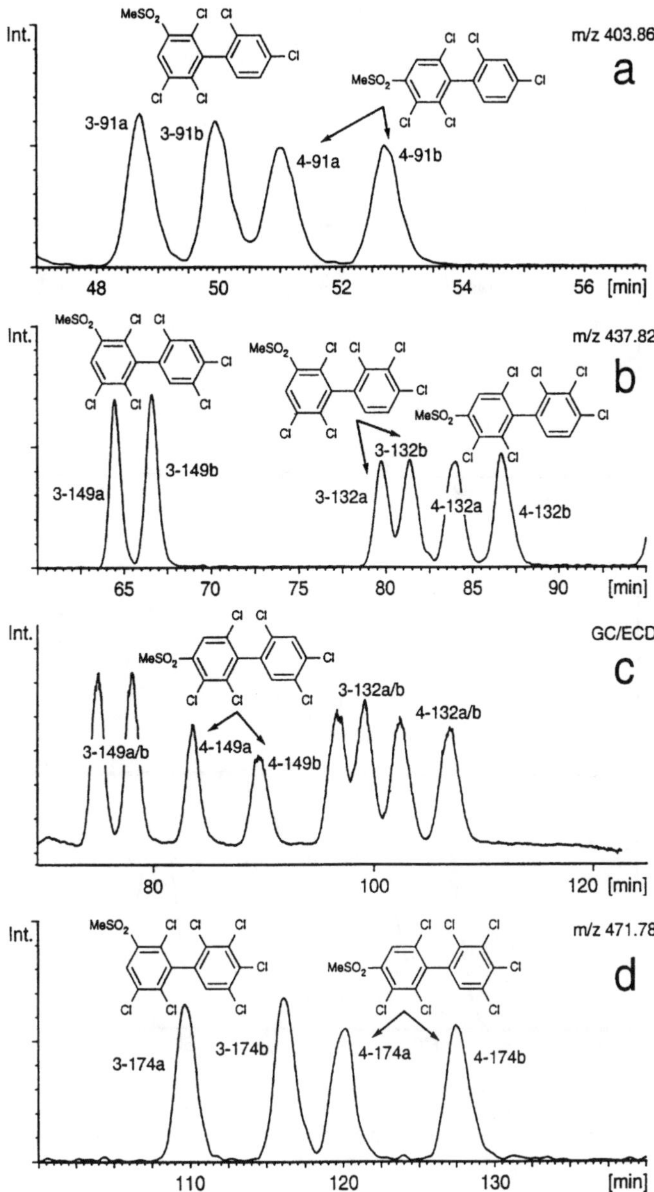

Fig. 3.22. Enantioselective cGC-MS ion chromatograms of seven MeSO$_2$-PCBs: 3–91 (*note: see important comment in the text*), 4–91, 3–149, 3–132, 4–132, 3–174 and 4–174 resolved into their enantiomers (a, b and d), and a cGC-ECD chromatogram with the enantiomers of 3–149, 4–149, 3–132 and 4–132 (c). The structural formulae and abbreviations of the sulfones are also given; from [35]

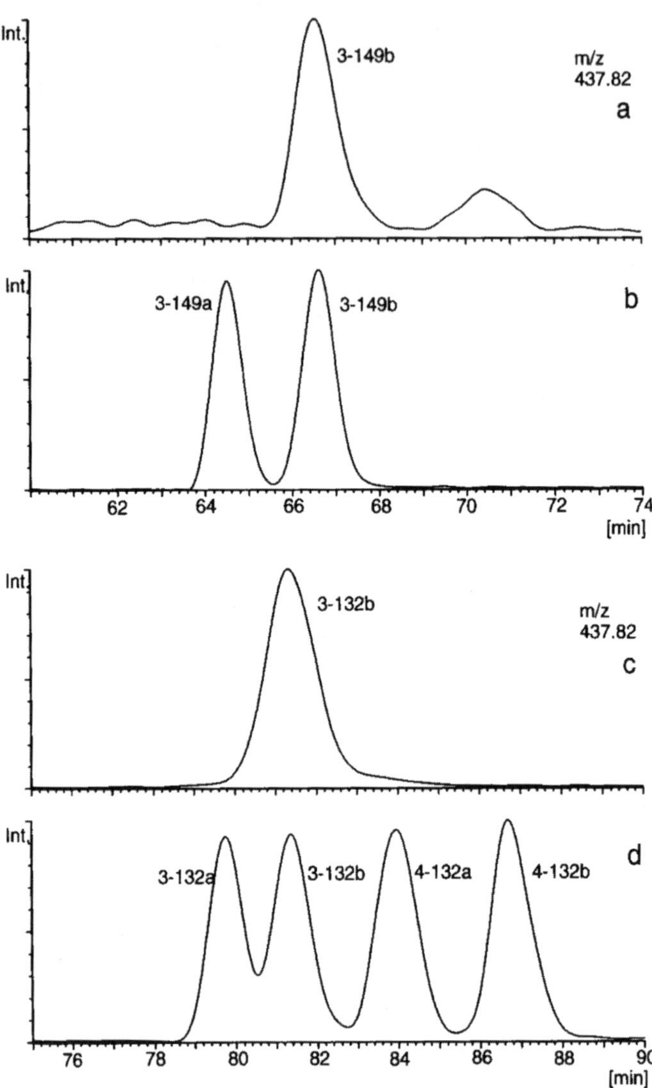

Fig. 3.23. SIR traces from HRMS of the racemic standard substances 3–149, 3–132 and 4–132 (b and d) and the corresponding SIR traces of a human liver sample extract (a and c); from [35]

an achiral column (SE 54), whereas for one sample cGC/MS full scan analysis was applied (see [35]). The enantioselective gas chromatographic analysis of tissue sample extracts with an ECD was not successful due to interferences with the complex liver matrix resulting in negative peaks in the chromatograms. Attempts to remove these interferences failed. The matrix problems were overcome by application of cGC/MS high-resolution SIR, which allowed a substantially more selective detection of the target compounds (Fig. 3.23).

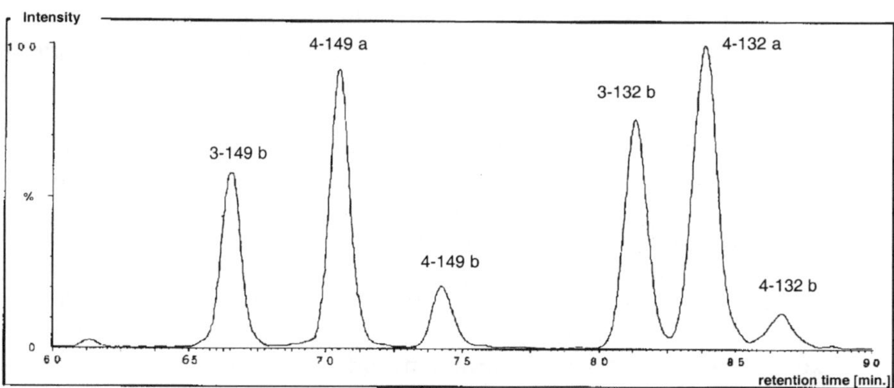

Fig. 3.24. Enantioselective separation of $MeSO_2$-hexaCBs in a rat liver sample two weeks after a single dose of the commercial PCB mixture Clophen A50; from [104]

In all the liver sample extracts investigated by Ellerichmann et al. [35], the second-eluting enantiomers of the $MeSO_2$-PCBs 3–149 and 3–132 were exclusively encountered, i.e., not only a high congener selective (as already previously observed), but also a high enantioselective liver retention of these $MeSO_2$-PCBs was determined. None of the $MeSO_2$-PCBs included in the study by Ellerichmann et al. were detected in the two human lung samples also analysed.

During the third period of the joint Swedish/German investigation, enantiomeric excesses of chiral methylsulfonyl PCBs in liver and adipose tissues from rats dosed with Clophen A50 were determined [104]. Rats were given one single oral dose of the commercial PCB product Clophen A50 dissolved in corn oil (25 mg/kg b.w.). The rats were sacrificed after 1, 2, 4, and 8 weeks and liver, lung, kidney, adrenal gland and fat were removed and kept frozen until the analysis.

In all samples analysed the concentrations of the 4-$MeSO_2$-CB91 and 4'-$MeSO_2$-CB132 were slightly higher than those of the 3-$MeSO_2$-PCB isomers in fat and liver. However, the ratio 4-$MeSO_2$-PCBs/3-$MeSO_2$-PCBs was still almost 1:1 for these $MeSO_2$-PCB congeners in both liver and fat. It is important to note that comparable enantiomeric ratios were found in fat and liver tissue extracts: in all samples analysed by Larsson et al. [104], only the second-eluting enantiomers of 3'-$MeSO_2$-CB132 and 3-$MeSO_2$-CB149 were present (Fig. 3.24). In contrast, 4-$MeSO_2$-CB91, 4'-$MeSO_2$-CB132 and 4-$MeSO_2$-CB149 were dominated by the first-eluting enantiomers, although minor amounts of the second-eluting enantiomers were also found. The ratios of the first- and second-eluting enantiomers (the quotient a/b) for 4-$MeSO_2$-CB91 were in the range 7–11, for 4-$MeSO_2$-CB149 2–8 and for 3'-$MeSO_2$-CB132 6–13. This indicates that either both enantiomers are formed or, if only one enantiomer is formed, that this enantiomer is converted to both optical forms. This is true at least for the 4-$MeSO_2$-PCBs. For the 3-$MeSO_2$-PCBs there were no indications of another enantiomer. Furthermore, it cannot be excluded that enantioselective transport processes may also play an additional role.

A surprising result was recently presented by Larsson et al. (unpublished, Poster at the DIOXIN'99 Symposium, Venice) with regard to the enantiomeric excesses of $MeSO_2$-PCBs in rat lung sample extracts, determined for the same animals from which the adipose and liver tissues had been analysed. It turned out that the second-eluting enantiomers of 4-$MeSO_2$-CB149 and 4-$MeSO_2$-CB132 were clearly more abundant than the first-eluting ones. This preliminary result has yet to be verified by the analysis of additional samples.

Alder et al. [72] found high concentrations of three chiral chlorobornanes in human milk from women with a high proportion of seafood in their diet. In addition, the same toxaphene components were determined in the adipose tissue of monkeys treated for one year with the whole complex mixture of toxaphene. In all extracts, two out of the three toxaphene indicator compounds could be detected and separated into the enantiomeric pairs, i.e., Parlar26 (B8–1413) and Parlar50 (B9–1679). Regarding Parlar26, the authors obtained enantiomeric ratios between 1.07 and 1.28 for human milk, while for monkey adipose tissue two values of 1.29 and 1.31 were determined. The ERs for Parlar50 in the same samples ranged from 1.06 to 1.33 in human milk and 1.35 to 1.44 in monkey adipose tissue. Alder et al. concluded that the first-eluting enantiomer of the two toxaphenes was more abundant in warm-blooded animals than in fish samples that were also included in their study (see Sect. 3.2.1.1).

Buser and Müller [24] applied enantioselective HRGC (PS086+10% permethylated β-cyclodextrin) with EI and ECNI-MS detection to the analysis of a human tissue extract, placing emphasis on oxychlordane and heptachlor *exo*-epoxide. The chromatograms showed a separation of oxychlordane into a pair of enantiomers with the first-eluting one being clearly more abundant. With regard to heptachlor *exo*-epoxide the author reported the notable result that in human adipose extracts also the first-eluting enantiomer dominated, while in aquatic species from the Baltic (herring, salmon, seal) the second-eluting enantiomer was more abundant

3.2.2
Enantioselective Permeation Through the Blood-Brain Barrier

During analyses of brain tissues from harbour seals (*Phoca vitulina* L.), Hühnerfuss et al. [105] and Möller et al. [19] encountered a surprising phenomenon. In brain samples from eight Icelandic seals almost exclusively (+)-α-HCH was found (Fig. 3.25). In contrast, blubber tissue from the same animals yielded enantiomeric ratios of only 1.2 to 1.4. For comparison, tissue samples from dead harbour seals found in the German Bight in 1989 were included in the study. The concentrations, as well as the enantiomeric ratios, of α-HCH determined in blubber and brain tissues from these three harbour seals and tissue samples from eight Icelandic harbour seals are summarised in Table 3.18.

This finding was especially interesting because of the widespread hypothesis that the blood-brain barrier is also partially effective towards certain fat-soluble organic pollutants. The "blood-brain barrier" is normally understood as mech-

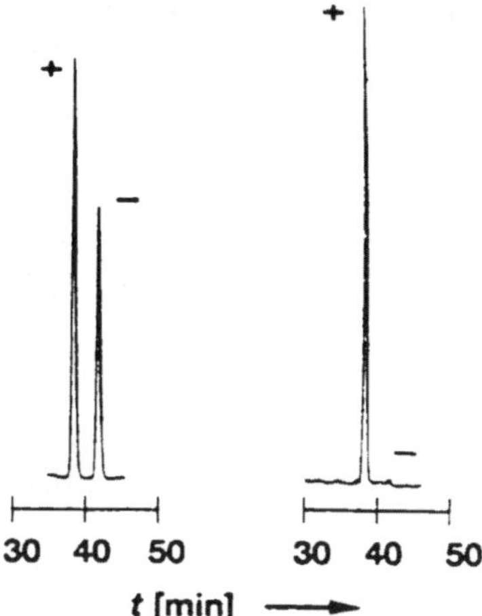

Fig. 3.25. Comparison of chromatograms [(+)- and (–)-α-HCH] for *n*-hexane extracts of blubber (*left*) and brain tissue (*right*) from the harbour seal (*Phoca vitulina* L.); from [19]

anisms which greatly inhibit the transport of *non*-lipid-soluble substances such as proteins from blood vessels into the surrounding interstitial nervous system tissue (glia) and the capillary endothelium. These mechanisms thus ensure a constant environment for the neurons. The respiratory gases carbon dioxide and oxygen, however, can cross the capillary walls with ease. The exact mechanisms involved, especially for the selective inhibition of entry for certain lipid-soluble pollutants into the brain tissue, still require more detailed study. Whether the barrier effect is enhanced by enzymes in the endothelial cells (enzymatic barrier) also still appears controversial.

Concentrations of polychlorinated biphenyls (PCBs) in brain have been found to be a factor of ten lower than those found in other organs of the same animals. This well-known effect is also often ascribed to the influence of the blood-brain barrier. Other authors, however, have shown that α-HCH is accumulated very efficiently in the brain tissue of mice and rats. Brain samples from rats have also revealed the first indications of preferential accumulation of (+)-α-HCH [108]. The systematic investigation of eight Icelandic seals carried out by Hühnerfuss and co-workers [19, 105], as well as independent investigations of brain tissues from a neonatal fur seal and a stillborn fur seal by Mössner et al. [41], have confirmed that this phenomenon can obviously be generalised, at least for the less pollutant-exposed Icelandic animals (Table 3.18). The three seals from the German Bight appear to have been subject to higher impact by α-HCH which is reflected by the higher concentrations in the brain tissue of these animals. Whether or not the lower enantiomeric ratios determined in the brain tissue extracts of the latter seals are a consequence of a partial break through the blood-brain barrier, perhaps due to the higher concentrations, must be left open, because three animals only were investigated, which is not sufficient for such a general statement.

Table 3.18. Enantiomeric ratios (ER) and concentrations of α-HCH determined in tissue samples of animals from the marine ecosystem (harbour seals, fur seals and Eider ducks); f≡female; m≡male; juv≡juvenile; ad≡adult; y≡years

Animal; *Sample Location*	Tissue	ER (+)/(−)	Conc. (µg/g lipid) (+)-α–HCH	(−)-α–HCH	Ref(s)
Harbour seal; *German Bight 1*	blubber	4.47	0.05	0.01	[105]
	spinal marrow	19.89	0.21	0.01	
Harbour seal; *German Bight 2*	blubber	1.54	0.03	0.02	[105]
	brain	16.34	0.22	0.01	
Harbour seal; *German Bight 3*	blubber	2.83	0.05	0.02	[105]
	brain	7.9	0.23	0.03	
Harbour seal; *Iceland 1*	blubber	1.36	0.006	0.004	[19, 105]
	brain	∞	0.08	–	
	spinal marrow	∞	0.05	–	
Harbour seal; *Iceland 2*	blubber	1.26	0.011	0.009	[19, 105]
	brain	∞	0.11	–	
	spinal marrow	99.54	0.04	–	
Harbour seal; *Iceland 3*	blubber	1.21	0.011	0.009	[19, 105]
	brain	∞	0.11	–	
	spinal marrow	∞	0.07	–	
Harbour seal; *Iceland 4*	brain	∞	0.09	–	[19, 105]
	spinal marrow	∞	0.06	–	
Harbour seal; *Iceland 5*	brain	66,24	0.08	–	[19, 105]
	spinal marrow	30.18	0.04	–	
Harbour seal; *Iceland 6*	brain	55.64	0.10	–	[19, 105]
	spinal marrow	∞	0.06	–	
Harbour seal; *Iceland 7*	brain	∞	0.06	–	[19, 105]
	spinal marrow	∞	0.03	–	
Harbour seal; *Iceland 8*	brain	∞	0.06	–	[19, 105]
	spinal marrow	22.84	0.03	–	
Neonatal fur seal; *Pribilof Islands, Alaska*	blubber	1.88/1.67			[41]
	liver	1.66/1.45/1.64	–	–	
	lung	1.55/1.47	–	–	
	brain	30.00/26.2/26.2	–	–	
Stillborn fur seal; *Pribilof Islands, Alaska*	blubber	1.85/1.60	–	–	[41]
	liver	1.83/1.64	–	–	
Eider duck (f, juv)	brain	∞	–	–	[106, 107]
Eider duck (f, juv)	brain	∞	–	–	
Eider duck (f, juv)	brain	∞	–	–	
Eider duck (m, 1 y)	brain	9.00	–	–	
Eider duck (m, 1 y)	brain	9.00	–	–	
Eider duck (f juv)	brain	9.00	–	–	
Eider duck (f, juv)	brain	5.67	–	–	
Eider duck (f, juv)	brain	3.35	–	–	
Eider duck (m, 2 y)	brain	2.33	–	–	
Eider duck (m, 2 y)	brain	2.23	–	–	

Table 3.18. (continued)

Animal; Sample Location	Tissue	ER (+)/(−)	Conc. (µg/g lipid) (+)-α–HCH	(−)-α–HCH	Ref(s)
Eider duck (f, juv)	brain	1.94	−	−	
Eider duck (m, 1 y)	brain	1.56	−	−	
Eider duck (f, juv)	brain	1.50	−	−	
Eider duck (m, 1 y)	brain	1.38	−	−	
Eider duck (m, 1 y)	brain	1.38	−	−	
Eider duck (f, ad)	brain	1.00	−	−	

Table 3.18 also summarises ER values of α-HCH for brain tissue extracts of common Eider ducks (*Somateria mollissima* L.) [106, 107]. Again, a preferential enrichment of the (+)-enantiomer in brain tissue can be inferred from these data, which may suggest that nearly exclusively (+)-α-HCH is able to penetrate the blood-brain barrier, while (−)-α-HCH is largely held back by it.

In a subsequent study, Möller et al. raised the question as to whether or not the preferential enrichment of (+)-α-HCH in the brain tissue of animals from the marine environment can be generalised, in particular, whether it is also valid for terrestrial biota of higher trophic level. As an example, the enrichment of α-HCH enantiomers has been investigated in fat, liver, and brain tissue samples of sheep bred in the northern German state, Schleswig-Holstein (see Sect. 3.2.1.2; Table 3.14; Fig. 3.18). The results are particularly interesting as, again in the brain tissues of seven sheep, enantiomeric ratios larger than one were encountered (ER 1.38–3.76); however, in the liver (ER 0.69–0.96) and fat tissues (ER 0.56–0.85) of the same animals values smaller than one were found. These results show that the enrichment of the α-HCH enantiomers in sheep fat and liver tissues is different from those determined in the respective tissues of marine biota of high trophic level. Möller et al. assumed that the variety of enzymatic systems encountered in the livers of sheep may give rise to very different transformation pathways. On the other hand, the enrichment of (+)-α-HCH in the brain in the sheep suggests that the blood-brain barrier, discussed above for marine biota, functions very similar in the case of sheep.

These very clear results encouraged Möller to expand the investigations to the analysis of human brain tissues [107]. A characterisation of the persons whose brain tissues had been supplied by the Department of "Neuropathologie des Universitätskrankenhauses Eppendorf" (University of Hamburg), their cause of death, as well as the analytical results, are summarised in Table 3.19. A separate determination of both the concentrations of α-HCH as well as of the (+)-α-HCH/(−)-α-HCH ratios was carried out for the cerebellum, white matter and grey matter tissues which exert different functions. Furthermore, it may be interesting to note that these three brain sections possess different lipid contents, i.e., in the cerebellum an average value of about 4.5%, in the white matter of 13.3%, and in the grey matter of 7.3% were found. These values have to be kept in mind when comparing the seemingly different concentrations in these three

Table 3.19. Characterisation of human brain samples, as well as α-HCH concentrations (μg/g EOM) and enantiomeric ratios (ER), determined in cerebellum, whiter matter and grey matter tissue extracts [107]

Sample No. *sex/age* *(cause of death)*	Cerebellum α-HCH-conc. (μg/g EOM)	ER (+)/(−)	White matter α-HCH-conc. (μg/g EOM)	ER (+)/(−)	Grey matter α-HCH-conc. (μg/g EOM)	ER (+)/(−)
1 *female/77* *(cardiac infarction)*	0.011	0.92	0.003	2.33	0.004	1.50
2 *male/58* *(Angina pectoris)*	0.010	0.79	0.010	0.75	0.004	0.79
3 *female/90* *(heart failure)*	0.013	0.67	0.008	0.96	0.016	0.92
4 *female/97* *(heart failure)*	0.012	0.59	0.012	4.00	0.005	2.23
5 *female/76* *(stomach cancer)*	0.026	0.75	0.010	1.86	0.011	1.50

brain sections. A normalisation to wet weight would largely supply comparable concentrations.

As shown in Table 3.19, high concentrations of α-HCH were determined in all five cerebellum samples, and in all five cases an enantiomeric ratio smaller than one was found, i.e., the (−)-enantiomer was more abundant than the (+)-enantiomer. This result was insofar very surprising as it is at variance with all other results determined in brain tissue extracts obtained from biota of different trophic levels, be it from the marine or terrestrial ecosystems. With regard to the enantiomeric ratios for the white and the grey matter samples, the situation appears to be more complex. From samples 2 and 3, in both cases ER values smaller than one were obtained, while for samples 1, 4 and 5, ER values larger than one were determined. Though the same enantiomeric ratios were determined for both brain sections, no clear correlations can be inferred, neither from sex nor from age, possibly because of the small data set. For the same reason it would also be too early to relate the high α-HCH concentrations encountered in all three brain sections of person #5 to the disease cancer. It cannot be excluded that a strong disease may reduce the function of the blood-brain barrier, similar to the results obtained for seals and common Eider ducks found dead along the coast of the German Bight. However, too many variables have to be considered which would require a much larger data set, before allowing a clearer assessment with regard to such correlations.

3.3
Transformation/Accumulation of Xenobiotics in Water, Sediment and Biota of Wastewater Treatment Plants

A most comprehensive study of the enantioselective transformation and accumulation of chiral pollutants in water, sediment and biota of a wastewater treatment plant has recently been published by Gatermann et al. [109, 111–113]. The authors placed emphasis on polycyclic musks and their main metabolites. Musk, one of the oldest-known ingredients of perfumes, is an odoriferous animal product that functions as a pheromone in nature, and it has also been isolated from plant products like Angelica root oil and Ambrette seed oil. It has a musky odour that is still considered essential in perfumery [114]. Twentieth-century chemical investigations and the subsequent laboratory synthesis of surrogates have permitted cheap synthetic chemicals to replace the animal products. In particular, musk xylene (MX) and musk ketone (MK) have been widely used as fragrance ingredients in soaps, laundry detergents and cosmetics. As a consequence, considerable concentrations of these nitro musks were and still are being found in municipal wastewater treatment plants. In 1996 the world-wide production rate of synthetic musk compounds amounted to 8000 tons, where polycyclic musks comprised 70% (mainly HHCB and AHTN, see below), nitro benzenoid musks 25% (mainly musk xylene and musk ketone), and macrocyclic musks 5%, respectively [115].

The occurrence of nitro musks in the aquatic environment was first reported in 1981 [116, 117]. Subsequently, these compounds were detected in marine and freshwater [118, 119], in fish, mussels and shrimps [119, 120], in human adipose tissue and in breast milk [121–123]. Due to its obviously high bioaccumulation potential, MX was banned in Japan in the 1980s, and in Europe MX and MK are nowadays under discussion because of their environmental and toxicological impact. In 1998, MX was included in the 'list of chemicals for priority action of the EU and OSPAR commission' [124, 125]. Critical considerations concerning analysis and bioaccumulation of musk xylene as well as other synthetic nitro musks are summarised in the literature [126]. Furthermore, it is well known that one of the major pathways of biological transformation of nitroaromatic compounds is the reduction of the nitro group to the corresponding amine. In a recent paper, Gatermann et al. and Rimkus discussed the occurrence of the MX and MK amine metabolites in the aquatic environment, in particular in wastewater treatment plants [115, 127].

In view of the increasing public concern about nitro musks and their metabolites in various environmental compartments, in Europe polycyclic (non-nitro) musk fragrances based on indane and tetraline play an increasing role [128]. In western Europe, HHCB (1,3,4,6,7,8-hexahydro-4,6,6,7,8,8-hexamethylcyclopenta[g]-2-benzopyran; Galaxolide®) as well as AHTN [1-(5,6,7,8-tetrahydro-3,5,5,6,8,8-hexamethyl-2-naphthalenyl)ethanone; Tonalide®] are of prominent importance, while in North America the nitro musks are still assumed to dominate. In Europe HHCB and AHTN are classified as 'high volume chemicals' with an average use per capita in 1995 of about 15.5 mg/day [129].

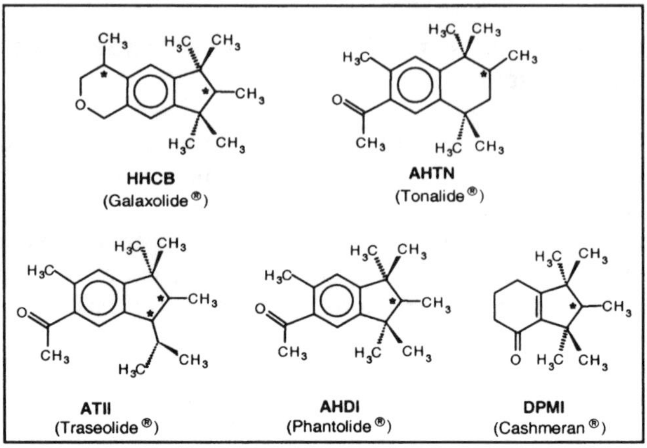

Fig. 3.26. Short names, trade names, and chemical structures of five polycyclic musk compounds [111]

Recently, Eschke et al. [130, 131] identified for the first time polycyclic musk compounds in water and fish from the River Ruhr and some municipal sewage treatment plants and, meanwhile, they have also been found in various environmental compartments [115], the environmental concentrations of the main pollutants HHCB and AHTN exceeding those of the nitro musks. Furthermore, polycyclic musks were also detected in human adipose tissue and human milk [123], which gave rise to a critical discussion with special emphasis on potential dermal resorption and chronic toxicity [128, 132, 133].

The synthetic polycyclic musks HHCB and AHTN, as well as ATII (1-[2,3-dihydro-1,1,2,6-tetramethyl-3-(1-methylethyl)-1H-inden-5-yl]ethanone; Traseolide®) and AHDI [1-(2,3-dihydro-1,1,2,3,3,6-hexamethyl-1H-inden-5-yl)ethanone; Phantolide®)], were included in studies recently published by Gatermann et al. [37, 112], Hühnerfuss et al. [111], Biselli et al. [113], and Franke et al. [134]. The stereochemical structures of these four polycyclic musk compounds can be found in Fig. 3.26. All derivatives are chiral compounds, where HHCB and ATII exhibit two asymmetric centres and thus two diastereomeric pairs of enantiomers. For HHCB the enantioselective syntheses and separations of all four stereoisomers were achieved, and thus it could be proved that the 4S,7R/S-isomers are the powerful, musky components [135]. Furthermore, the authors postulate a high three-dimensional structural similarity between 5α-androst-16-en-3-one and these two active isomers.

3.3.1
Separation of the Standard Substances of Polycyclic Musks

In the case of HHCB, the separation of the diastereomers on capillary columns commonly used in pesticide and PCB residue analysis turned out to be a chal-

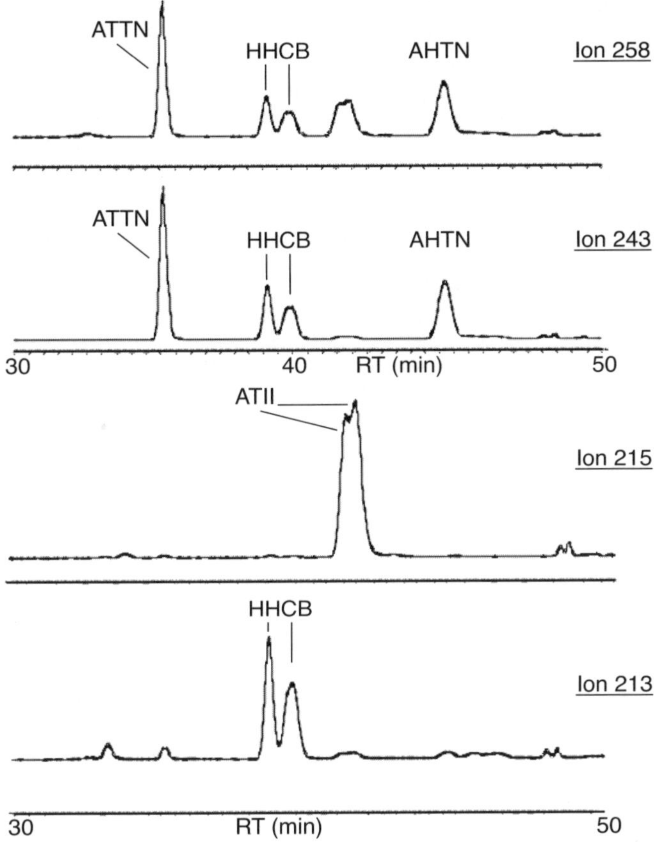

Fig. 3.27. SIM fragmentograms of a standard mixture containing HHCB, AHTN, and ATII (1 ng/µL each), stationary phase: 1:4 mixture of OV-1701/octakis(2,3,6-tri-O-ethyl)-γ-cyclodextrin [37]

lenge. Only in two studies the diastereomers were separated, i.e., in extracts of human adipose tissue and in fish extracts using a methylpolysiloxane phase with 12–15% phenyl groups and a polyethylene glycol phase, respectively [115]. The successful enantioselective cGC separations of four of the polycyclic musks shown in Fig. 3.26, including the diastereomeric pairs of enantiomers, are reported in the literature [37, 111–113, 134]. In the investigation carried out by Gatermann et al. [37, 112], two different cyclodextrin-type chiral stationary phases were tested. Using a column coated with a 1:4 mixture of OV-1701/octakis(2,3,6-tri-O-ethyl)-γ-cyclodextrin, only a baseline separation of the HHCB diastereomers was achieved, where *cis*-HHCB was the first-eluting isomer. The enantiomers of ATII were partly separated, whereas no separation of the AHTN enantiomers was obtained (Fig. 3.27). Accordingly, this column can only be used for a determination of the *cis/trans* ratio of HHCB.

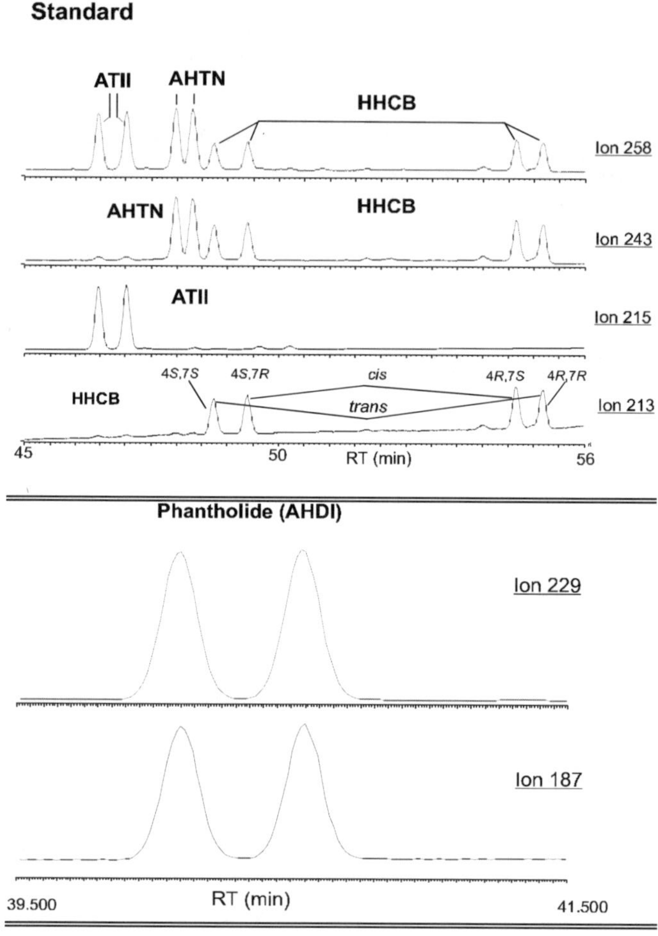

Fig. 3.28. SIM fragmentograms of a standard mixture containing HHCB, AHTN, ATII, and AHDI (1 ng/µL each), stationary phase: 1:1 mixture of OV-1701/heptakis(6-O-*tert*-butyl-dimethylsilyl-2,3-di-O-methyl)-β-cyclodextrin [37]

A satisfactory enantiomer separation of all chiral polycyclic musks was achieved with a column coated with a 1:1 mixture of OV-1701/heptakis(6-O-*tert*-butyldimethylsilyl-2,3-di-O-methyl)-β-cyclodextrin (TBDMS-CD). Selected ion monitoring (SIM) fragmentograms (*m/z* 187, 213, 215, 229, 243, and 258) of a standard mixture containing HHCB, AHTN, ATII, and AHDI (1 ng/µL each) are shown in Fig. 3.28. As the technical formulation of HHCB used in fragrance formulations contains different chiral by-products with the same fragmentation ions (*m/z* 258 and 243 [136]), possible coelution of these by-products and the polycyclic musks themselves cannot be basically excluded. However, injection of the authentic mixture confirmed that no coelution between HHCB and the by-products takes place on this column (Fig. 3.28).

The assignment of the peaks to the diastereomeric pairs of HHCB was achieved by comparing the retention times of the authentic separated stereoisomers (see [113]) with those of the different isomers in the racemic standard. The two diastereomers responsible for the significant musky odour (4S configuration) elute first on the present β-cyclodextrin column. Thus it turned out that the technical HHCB consists of an approximately 1:1 mixture of both diastereomers with a slight excess of the *cis*-isomer. Since no significant changes in the *cis/trans* ratios in the water, sediment, sewage sludge and semipermeable membrane device (SP-MD) samples [109] were found, the physicochemical properties of the isomers are expected to be relatively similar. By way of contrast, the technical ATII contains more than 95% of the *trans*-isomer, presumably reflecting larger differences in the physicochemical properties of the diastereomers, which in turn results in a high surplus of this *trans*-diastereomer during the synthesis. In line with these assumptions, the chromatographic separations of HHCB and ATII on the chiral column are different. The ATII enantiomers of each diastereomeric pair elute closely together, whereas the *cis*- and *trans*-isomers were widely separated. However, for HHCB the 4S- and the 4R-diastereomers eluted closely together, whereas the enantiomers were extremely widely separated (see Fig. 3.28). In addition, the separation of the enantiomers of AHDI is also shown in Fig. 3.28 (*m/z* 229 and 187).

3.3.2
Sampling Locations and Samples

The experimental site of the study carried out by Gatermann et al. was a sewage treatment plant in the Federal State of Schleswig-Holstein (Germany). At first, the sewage enters the treatment plant, and, subsequently, the treated wastewater flows towards a pond, where it is allowed to remain for some weeks, assuming an average water exchange rate of about 17,000 to 22,000 m^3/d and taking into account an area of 128,000 m^2 and a depth of 4 m. The effluent of the pond is closed with bars thus enabling small fish only to enter or to escape from the pond. Therefore, equilibrium conditions can be assumed for synthetic musks in fish tissues of larger animals that have to remain within the pond. The fish population status can be characterised as 'good and stable', and no commercial and private fishing activities are allowed. Samples were taken as follows:
- at the effluent of the sewage plant, two 10 L 'discrete water samples' on June 24th and July 29th, 1997, and one 3 L 'bulked composite water sample' (i.e., taken throughout a 24 h cycle) on July 29th, 1997;
- a 50 g 'discrete sludge sample' on July 29th, 1997;
- from the pond of the sewage plant two 10 L 'discrete water samples' on June 24th and July 29th, 1997, respectively, two parallel 50 g sediment samples on May 5th, 1997. Close to this sampling site two series of SPMDs consisting of six samples each were exposed and taken after an exposition of 7 weeks (June 24th, 1997) and 12 weeks (July 29th, 1997), respectively.
- from the pond of the sewage plant 18 fish samples and one pooled mussel sample (Table 3.20). The fish were caught by electro-fishing on the same day,

Table 3.20. Concentrations (µg/g lipid) and enantiomeric ratios (ER) of HHCB, *trans*- and *cis*-HHCB, AHTN, *trans*-ATII, and AHDI in biota caught in the pond of a municipal wastewater treatment plant in summer 1997 [37]

Species	HHCB Conc.	*trans*-HHCB ER	*cis*-HHCB ER	AHTN Conc.	ER	*trans*-ATII Conc.	ER	AHDI Conc.	ER
Rudd	6.2	0.66	1.24	5.0	0.94	0.5	0.99	0.3	0.90
Rudd	7.1	0.55	1.10	5.7	0.91	0.5	0.98	0.3	1.12
Rudd	7.5	0.57	1.07	6.1	0.80	0.5	0.97	0.3	0.95
Tench	150	1.06	0.98	30	1.65	2.3	0.60	2.2	1.31
Tench	160	1.10	0.95	32	1.97	2.0	0.38	1.9	1.47
Tench, liver	72	1.01	0.85	16	1.98	1.1	0.33	0.9	1.78
Tench	160	0.88	1.00	35	1.74	2.2	0.51	2.1	1.30
Tench	150	1.03	0.98	42	1.65	2.6	0.46	2.5	1.28
Crucian carp	39	0.12	0.44	26	1.05	1.2	0.05	1.8	0.65
Crucian carp	59	0.12	0.42	31	1.24	1.5	0.08	2.7	0.81
Crucian carp	91	0.19	0.54	31	1.19	1.9	0.20	3.1	0.94
Crucian carp	71	0.14	0.48	30	1.15	1.4	0.10	2.5	0.78
Crucian carp	50	0.10	0.41	34	1.19	1.6	0.06	2.4	0.71
Crucian carp	66	0.10	0.46	40	1.14	1.5	0.05	2.3	0.66
Crucian carp	84	0.13	0.51	34	1.15	1.9	0.07	3.0	0.64
Crucian carp, liver	69	0.10	0.48	31	1.29	1.6	0.06	2.5	0.63
Eel	4.8	0.98	1.27	2.6	0.87	0.3	1.95	0.2	0.73
Eel	4.6	0.79	1.15	2.7		0.3	1.50	0.2	0.48
Zebra mussel	120	0.91	1.26	45	0.90	3.4	0.67	2.3	1.02

June 24th, 1997. Both the tench (*Tinca tinca* L.) and the crucian carp (*Carassius carassius* L.) tissue samples originated from a single fish, whereas the rudd (*Scardinius erythrophthalmus* L.), eel (*Anguilla anguilla* L.), and mussel samples were pooled. The mussel samples (*Dreissena polymorpha* PAL.) were collected from an uncontaminated lake ('Lake Gartow') near the city of Gartow, State of Lower Saxony (Germany), and were exposed to the pond within polypropylene nets next to the exposition place of the SPMDs.

For further information about the experimental details concerning sample collection, extraction and clean-up, separation and quantification as well as apparatus, chemicals and reagents used the reader is referred to the original paper [109]. Additional aspects with regard to quality assurance, especially to prevent laboratory contamination, information about recovery rates, variation tests and limits of quantification for the present analytical methods are also given therein.

3.3.3
Enantiomeric Ratios of ATII in Water and SPMD Samples

In the water and SPMD samples collected in the pond deviations from the racemic enantiomeric ratios (ERs) were found only for *trans*-ATII (in water samples: ER=0.7–0.8; in SPMD samples: ER=0.8), which is a strong pointer that pronounced enantioselective degradation of this specific polycyclic musk derivative takes place in the water of the sewage treatment plant [37]. This appears not to be the case for the other polycyclic musks investigated by Gatermann et al.

3.3.4
Enantiomeric Ratios of HHCB, ATII, AHTN, and AHDI in Biota Samples

Selected ion monitoring (SIM) fragmentograms of biota sample extracts, i.e., rudd, tench, crucian carp and Zebra mussel, are shown in Figs. 3.29 and 3.30. Concentrations and enantiomeric ratios of HHCB, *cis-* and *trans-*HHCB, AHTN,

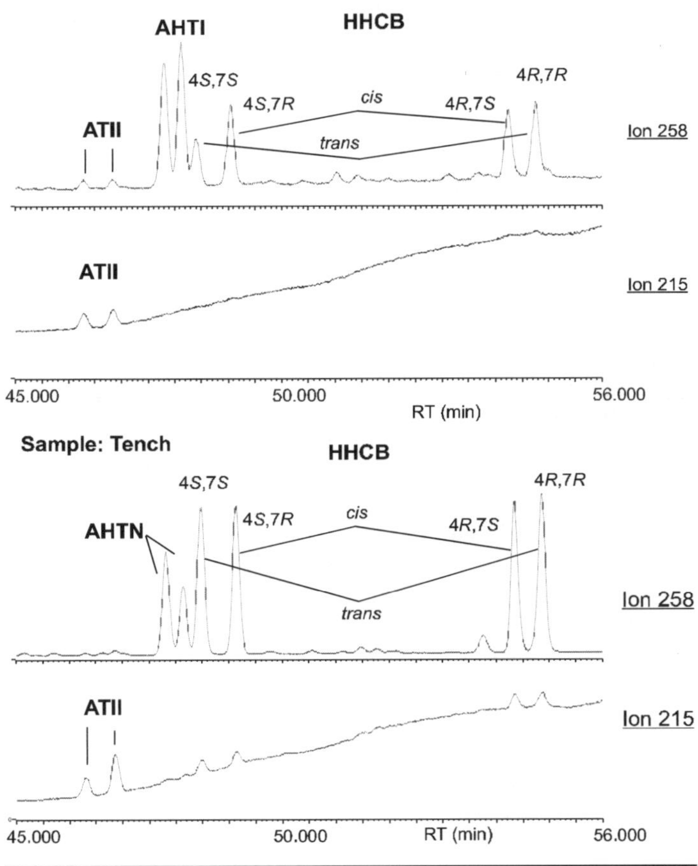

Fig. 3.29. SIM fragmentograms of rudd and tench sample extracts, stationary phase: 1:1 mixture of OV-1701/heptakis(6-*O-tert*-butyldimethylsilyl-2,3-di-*O*-methyl)-*β*-cyclodextrin [37]

trans-ATII, and AHDI for 18 fish samples and one pooled mussel sample from the pond of the sewage plant are summarised in Table 3.20. In the case of one tench and one crucian carp, liver samples were analysed in addition to the muscle tissue. The concentrations of HHCB and AHTN are discussed in [109]. Both the concentrations and the enantiomeric ratios appear to exhibit a species dependency. Clearly distinguishable clusters can be inferred from the data set. For example, for HHCB, the following species-dependent clusters can be stated: rudd (conc=6.2–7.5 µg/g lipid; ER_{trans}=0.55–0.66; ER_{cis}=1.07–1.24), tench (conc=150–160 µg/g lipid; ER_{trans}=0.88–1.10; ER_{cis}=0.95–1.00), crucian carp (conc=39–91 µg/g lipid; ER_{trans}=0.10–0.19; ER_{cis}=0.41–0.54), eel (conc=4.6–4.8 µg/g lipid; ER_{trans}=0.79–

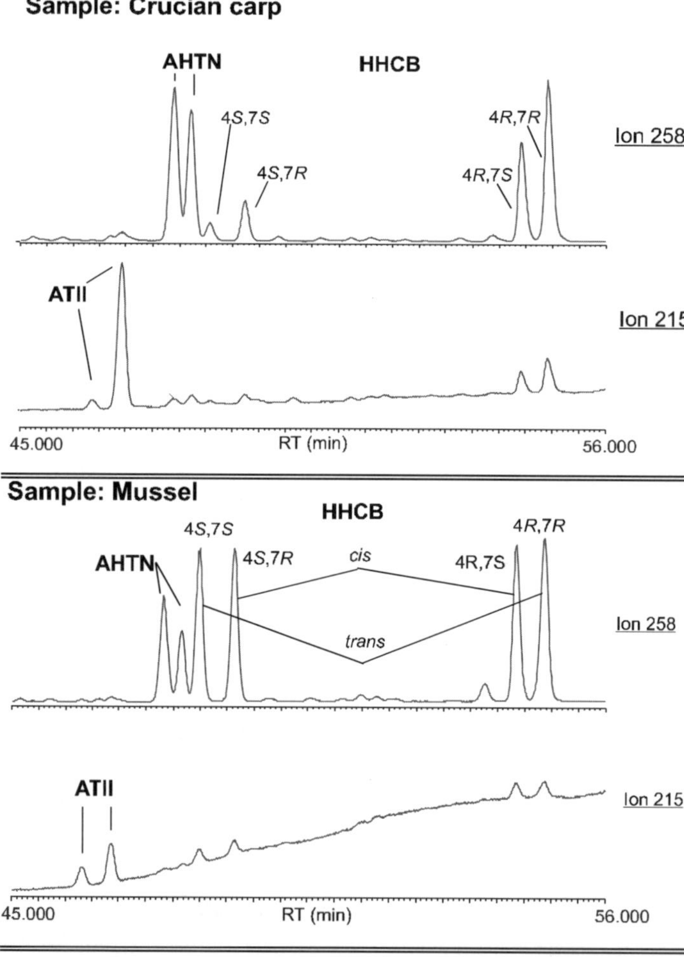

Fig. 3.30. SIM fragmentograms of crucian carp and Zebra mussel sample extracts, stationary phase: 1:1 mixture of OV-1701/heptakis(6-*O-tert*-butyldimethylsilyl-2,3-di-*O*-methyl)-*β*-cyclodextrin [37]

0.98; $ER_{cis}=1.15–1.27$), and mussel (conc=120 µg/g lipid; $ER_{trans}=0.91$; $ER_{cis}=1.26$). Similar clusters can be inferred from Table 3.20 for concentrations and enantiomeric ratios for AHTN, *trans*-ATII, and AHDI, respectively.

It is tentatively assumed that high concentrations and enantiomeric ratios close to racemic, i.e., $ER \approx 1.0$, indicate a low metabolisation potential of a species for the respective polycyclic musk derivative. On the other hand, caution has to be applied when inferring metabolisation potentials exclusively from lipid-based concentrations: low concentrations may reflect a stronger metabolisation, but it cannot be excluded that specific matrix effects, e.g., high lipid contents like in eel, may mimic a stronger metabolisation than actually encountered. An unequivocal parameter for an enantioselective transformation process, however, are ER values clearly different from one. On the basis of these assumptions, rudd appear to exhibit a strong enantioselective metabolisation potential for *trans*-HHCB.

Tench show a low enantioselective metabolisation potential for *trans*- and *cis*-HHCB, and a moderate one for AHTN and AHDI. Very strong enantioselective metabolisation can be concluded for *trans*- and *cis*-HHCB as well as for *trans*-ATII in crucian carp, while for AHDI a moderate metabolisation was observed. With ER values ≤ 0.1 for *trans*-HHCB and *trans*-ATII, the highest enantioselectivity in the study by Gatermann et al. was observed. In eel the high lipid content gives rise to low lipid-normalised concentrations, but the ER values indicate low to moderate enantioselective metabolisation for *trans*- and *cis*-HHCB and AHTN, while for *trans*-ATII and AHDI a stronger enantioselective metabolisation capability was found. For a more detailed discussion of the results obtained herein for crucian carp and tench, the reader should refer to Section 3.3.5.

The values for the pooled Zebra mussel sample, known to have a low metabolisation capability, reflect the water values, which were also determined in the study carried out by Gatermann et al. [37]. By way of contrast, in the liver the metabolic capacity is higher than in other organs. As a consequence, lower concentrations of the polycyclic musks as well as larger deviations from racemic were determined in the tench and crucian carp liver extracts (with the exception of *trans*-HHCB in the tench sample) as compared with the corresponding muscle tissues (Table 3.20).

As stated above, the lipid-based concentrations for the assignment of metabolisation potentials may be misleading. However, in those cases where both low concentrations and significant deviations of the enantiomeric ratios from one are encountered, it is justified to assume strong metabolisation capacities. This assumption is in line with the previous conclusions drawn by Wiberg et al. from enantioselective analyses of chlordanes in different fish species [46].

3.3.5
Comparison of Tench and Crucian Carp Samples

In order to support the conclusion that the lower concentrations encountered in crucian carp are caused by enantioselective degradation, theoretical concentra-

Table 3.21. Experimental and calculated concentrations of HHCB, AHTN, and ATII (µg/g lipid) [37]

Species	HHCB conc.		AHTN conc.		*trans*-ATII conc.	
	Experim.	Calc.	Experim.	Calc.	Experim.	Calc.
Tench	152	–	35	44	2.3	2.8
Tench, liver	72	–	16	21	1.1	1.7
Crucian carp	66	122	32	35	1.6	3.0
Crucian carp, liver	69	135	31	35	1.6	3.0
Zebra mussel	120	–	45	–	3.4	–

tions for passive bioconcentration were calculated by Gatermann et al. [37] on the following basis: it was tentatively assumed that only one enantiomer is transformed and that the prevailing enantiomer represents 50% of a "theoretical concentration" without metabolisation. In the case of *trans*-ATII, an ER of 0.8 (found in the water samples) instead of a racemic distribution was used as the basic value. The calculated values for each species as well as mean experimental values are summarised in Table 3.21. In the case of AHTN, the measured values in crucian carp and tench do not differ significantly and are comparable with the values in mussels, which implies that the concentration levels are explainable with passive bioconcentration only and no or little metabolisation. On the other hand, in crucian carp significantly lower concentrations of HHCB and ATII compared with tench as well as lower experimental than calculated values indicate strong enantioselective transformation of the first eluting *trans*-ATII enantiomer in crucian carp. In the case of HHCB, the calculation is more sophisticated, because the measured concentration is the sum of two diastereomers, *cis*- and *trans*-HHCB. The comparison of the stereochemical patterns in the crucian carp sample extracts and in standard solutions shows a different behaviour of the *cis*- and the *trans*-isomer: For the *cis*-isomer both enantiomers seem to be degraded compared with the latest-eluting enantiomer of *trans*-HHCB. Therefore, it was decided to choose this stereoisomer (4*R*,7*R*) as the basic value for the calculation. The mean calculated value (122 µg/g lipid) is as high as the mussel value and only slightly lower than the mean value of the tench samples (152 µg/g lipid). Thus, it can be concluded that for HHCB enantioselective transformation is also the most important process resulting in lower concentrations for crucian carp. In addition, a higher selectivity of the *trans*- compared with the *cis*-isomer can be assumed.

3.3.6
Implications of the Observed Enantioselectivity for Risk Assessment Studies

Furthermore, the in part high enantioselectivity in the metabolic capacity for the polycyclic musks and the well-defined binding activities of the polycyclic musks to olfactory receptors [137–140], the known effect of natural musk as pheromones [141–143] and the structural relationship, especially of HHCB, to steroid hormones [135] raise the question as to whether or not synthetic musks have an effect on the chemical communication and/or the hormone system of species in the aquatic and/or the terrestrial environment. However, this question can only be answered by systematic experiments in which tests with pure enantiomers have to be included. These enantiomers have recently become available by successful enantioselective HPLC separations [113]. Presently, all answers to this problem must remain speculative. If it is tentatively assumed that *Pfeiffer's rule*, which plays an important role in the prediction of drug efficiency [150, 174], can be transferred and adopted to the needs of environmental chiral assessments, a new approach for enantioselective environmental processes could be postulated: Pfeiffer conjectured that the better the drug-receptor match, the greater the drug potency and the higher the enantiomeric ratio (ER). Based on this general postulate, a new theory for enantiomer selective hazardous potency can be formulated: "*The better the diastereomeric complex built by enantiomer and receptor match, the greater is the environmental hazardous potency and the higher the enantiomer ratio of the chiral compound in environmental samples*". However, we would like to stress that this postulate is highly speculative and has yet to be verified.

3.3.7
A New Challenge: HHCB Metabolites

A comprehensive risk assessment study should also include the metabolites of the respective xenobiotics. In the case of the polycyclic musks, recent publications by Biselli [173] and Franke et al. [134] presented first results on HHCB metabolites, including their syntheses and their enantioselective gas chromatographic separations.

The main HHCB metabolite, 1,3,4,6,7,8-hexahydro-4,6,6,7,8,8-hexamethylcyclopenta[g]-2-benzopyran-1-one (hereafter "galaxolidone"), is shown in Fig. 3.31. Franke et al. [134] synthesised the racemic standard compound by oxidation of racemic HHCB using a finely powdered mixture of potassium permanganate and copper sulfate pentahydrate. Thus, they were for the first time

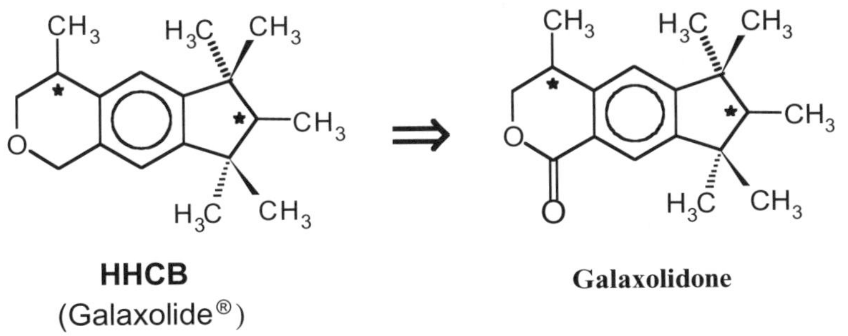

Fig. 3.31. Transformation of HHCB to galaxolidone

able to verify the presence of this compound in environmental samples. A re-evaluation of existing cGC/MS analyses showed that this metabolite is very common in surface waters including water from the Rivers Odra and Elbe. Franke et al. conjectured that galaxolidone is likely to be formed by autoxidation of the benzylic methylene group of HHCB, and so its occurrence in aquatic environmental samples can be explained by an abiotic process. Enzymatic oxidation at the benzylic position of HHCB may also occur, but the formation of galaxolidone is certainly not restricted to biotransformation reactions. For example, in laboratory experiments, Itrich et al. observed the oxidation of HHCB to the lactone in activated sewage sludge (i.e., biotransformation) and in abiotic controls [110].

Extensive enantioselective cGC/MS investigations of galaxolidone in different environmental compartments including different fish species and sediment have recently been reported by Biselli et al. [173].

3.3.8
Other Enantioselective Studies Including Sewage Water or Sludge

Isomer and enantioselective transformation processes of xenobiotics in sewage sludge were studied by Buser and Müller [7, 144], placing emphasis on the behaviour of the four most common HCH isomers (α-, β-, γ- and δ-HCH) in sewage sludge from the anaerobic digestor of a typical Swiss wastewater treatment plant. Approximately 250 g of sewage sludge in a 300 mL clear glass serum bottle was fortified with 100 μL of an ethyl ethanoate solution containing 400–500 μg of *rac*-α-HCH (experiment S1, see Table 3.22) or a technical HCH mixture (experiment S3). The fortification levels were below the water solubility limits of these compounds. A control experiment (S4) with sterilised sewage sludge was carried out with technical HCH added at the same concentration levels as above. A more detailed description of the experiments can be found in [7].

Initial experiments with 10-day exposures showed significant degradation of α- and γ-HCH in active sewage sludge to levels of <10 and <1%, respectively, and, to a lesser degree, of δ- and β-HCH. Furthermore, high enantioselectivity in the transformation of α-HCH was indicated by different degradation rates for the

Table 3.22. Degradation rate constants k (10^{-3} h^{-1}) and half-lives (τ) of the α-HCH enantiomers in sewage sludge; after [7]

Experiment	(+)-α-HCH	(−)-α-HCH
S1	20.2	7.26
S3	19.5	6.76
S4, sterilised	1.0	1.0
ratio $k_{\text{biotic}}/k_{\text{abiotic}}$	18.8	6.0
% biotic	95	86
τ (h), calculated from experiments S1, S3	35±0.5	99±3.5

(+)- and the (–)-enantiomers, resulting in an apparent enrichment of the (–)-α-HCH in the digested samples. By assuming pseudo-first-order kinetics, Buser and Müller inferred degradation rate constants k from their data. The rate constants showed good agreement for (+)- and (–)-α-HCH when incubated either as the pure isomers (experiment S1) or the technical product (S3), indicating no influence of one isomer on the degradation of another. The k values listed in Table 3.22 indicate a 2–3-fold difference between the two enantiomers of α-HCH, with (+)-α-HCH degrading faster. After incubation for 172 h, the initial ERs of 1.00 had changed to 0.14 and 0.12 for technical HCH and α-HCH, respectively, indicating a 7–8-fold excess of (–)-α-HCH. The k values for these degradations in sewage sludge were in the order γ-HCH >(+)-α-HCH >(–)-α-HCH >δ-HCH >β-HCH. As it is well known that the first transformation step consists of a *trans*-elimination of HCl [1, 2], it is not surprising that the degradation rates determined by Buser and Müller decrease with decreasing numbers of axial Cl substituents. In sterilised sludge (experiment S4), the degradation rates of all HCHs were 5 to 20 times lower than those in active sludge.

Furthermore, it is interesting to note that Buser and Müller observed a conversion of γ-HCH into α-HCH, similar to the isomerisation described by Ludwig et al. for the microbial degradation of γ-HCH [1, 5]. In the case of the sewage sludge experiment, the amount of α-HCH increased from 0.05 to about 0.5% after 50 h of incubation and then decreased again. The ER of approx. 0.8 is interpreted by Buser and Müller as the result of a nonenantioselective isomerisation of γ-HCH into α-HCH followed by a subsequent faster transformation of (+)-α-HCH and thus an enantioselective process. In addition, the authors note an absence of potential initial metabolites in anaerobic degradation on HCHs, in particular, the pentachlorocyclohexenes. Therefore, they suggest that these metabolites were further degraded at rates that are significantly faster than the rate of their formation. This conclusion is supported by earlier laboratory measurements on microbial transformation of HCHs and their metabolites carried out by Ludwig et al. [1, 5].

Recent investigations by Weigel have shown that drugs are being found in the aquatic environment in larger concentrations than previously expected [168]. It can be safely assumed that the larger part of these pharmaceutical compounds reached the environment via wastewater treatment plants. This hypothesis was supported by Buser et al. [169], who studied the occurrence and environmental behaviour of the chiral drug ibuprofen (2-[4-(2-methyl)propyl]phenylpropanoic acid; Fig. 3.32) in surface waters and in wastewater. Ibuprofen is a nonsteroidal antiinflammatory (NSAID), analgesic, and antipyretic drug widely used in the treatment of rheumatic disorders, pain, and fever. It exhibits an estimated annual global production of several kilotons, and, according to Buser et al. [169], it is the third-most popular drug in the world. It is an important nonprescription drug requiring a relatively high therapeutic dose (600–1200 mg/d). It is excreted to a significant degree (70–80% of the therapeutic dose) as the parent compound (free or conjugated) or in the form of metabolites. Its physicochemical properties suggest a rather high mobility in the aquatic environment, and, in fact, sev-

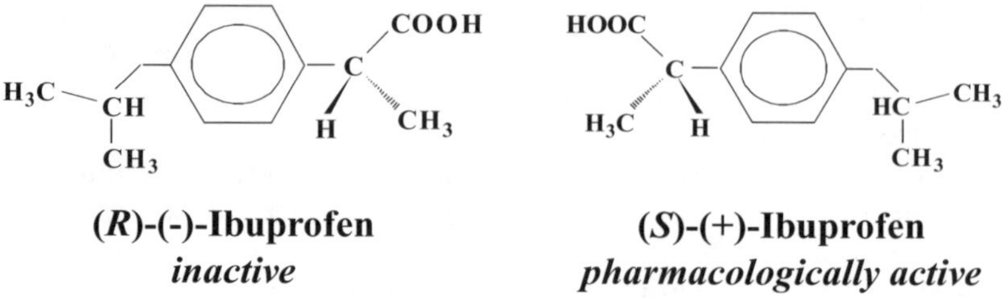

(R)-(-)-Ibuprofen
inactive

(S)-(+)-Ibuprofen
pharmacologically active

Fig. 3.32. Enantiomers of the antiinflammatory, analgesic, and antipyretic drug ibuprofen, i.e. 2-[4-(2-methyl)propyl]phenylpropanoic acid

eral research groups, such as those of Ternes [170] and Buser [169], have detected ibuprofen in wastewater and in rivers along with several other pharmaceutical compounds. There appears to be growing concern on the occurrence, fate, and possible effects of such substances in the environment [55]. For example, Ternes studied the occurrence of 32 drug residues belonging to different medicinal classes such as antiphlogistics, lipid regulators, psychiatric drugs, anti-epileptic drugs, betablockers and β_2-sympathomimetics as well as five metabolites in German municipal sewage treatment plant (STP) discharges, river water and stream water [170]. Due to the incomplete removal of drug residues from wastewater, >80% of the selected drugs were detectable in municipal STP effluent with concentrations of up to 6.3 µg/L (carbamazepine), thus resulting in the contamination of the receiving waters. Twenty different drugs and four corresponding metabolites were measured in river and stream waters by Ternes. Mainly acidic drugs, such as the lipid regulators bezafibrate, gemfibrozil, antiphlogistics diclofenac, ibuprofen, indometacine, naproxen, phenazone and metabolites, clofibric acid, fenofibric acid and salicylic acid as well as neutral or weak basic drugs such as betablockers metoprolol, propranolol and the antiepileptic drug carbamazepine, were present in the rivers and streams, mostly in the ng/L range. However, maximum concentrations attained values of 3.1 µg/L and medians as high as 0.35 µg/L (both bezafibrate). The drugs detected in the environment were predominantly applied in human medicine. In conclusion, several chiral pharmaceutical compounds enter the environment, thus offering a new field of enantioselective analysis of their fate and of their toxicological impact on aquatic and terrestrial biota.

A very comprehensive example for this new approach was recently published by Buser et al. [169], who collected water samples from wastewater treatment plants located in Gossau, Pfäffikon and Uster (Switzerland; all samples 24 h flow proportionally collected) of influent to the biological stage (raw sewage) and treated effluent emitted to the rivers. Furthermore, water samples from several Swiss lakes and rivers were analysed. The study also included incubation experiments with lake water and with activated sewage sludge as well as characterisation of urinary metabolites. Ibuprofen was present in influents of the wastewater treatment plants at concentrations of up to 3 µg/L with a high enantiomeric excess of the pharmacologically active S-enantiomer (see Fig. 3.32; $ER_{S/R} = 5.5-8$), as from human urinary excretion ($ER_{S/R}=19$). The principal human urinary metabolites of ibuprofen, hydroxy-ibuprofen and carboxyl-ibuprofen, were observed in the wastewater influents at even higher concentrations. In contrast to other pharmaceutical compounds such as clofibric acid and diclofenac, ibuprofen and its metabolites are then efficiently transformed (>95%) during treatment in wastewater treatment plants. The residual ibuprofen in the effluents showed a lower enantiomeric excess of the S-enantiomer ($ER_{S/R}=0.9$–2), indicating that the S-enantiomer is somewhat faster transformed than (R)-ibuprofen. In rivers and lakes, ibuprofen was detected at concentrations of up to 8 ng/L, generally with some excess of the S-enantiomer ($ER_{S/R}$ up to 2.0). The laboratory incubation experiments confirmed the rapid transformation of ibuprofen. Incubation of ibuprofen in influent water mixed with activated sludge indicated some initial phase with little or no dissipation, followed by rapid dissipation of ibuprofen and its principal metabolites to levels <1–3% after 8 h. In addition, the enantiomeric composition of ibuprofen changed during incubation from an initial $ER_{S/R}$ of 5.7 to an $ER_{S/R}$ of 2.7. The rapid and almost complete dissipation points to biologically mediated transformation rather than other processes such as sorption or uptake by sludge. The faster transformation of (S)-ibuprofen observed by Buser et al. is consistent with their results from the wastewater samples where also a trend to lower ERs in the effluents was determined. Incubation of lake water fortified with *rac*-ibuprofen indicated a faster dissipation of the S-enantiomer, thus resulting eventually in residues with a reversed ($R>S$) enantiomer composition as compared with that from human metabolism. The data from the sterile control sample showed no transformation of ibuprofen for up to 37 d, even when exposed to daylight. However, ibuprofen was transformed under nonsterile conditions in daylight as well as in the dark. Under both conditions a faster transformation of the S-enantiomer was observed, leading to an excess of (R)-ibuprofen in the residues (dark: $ER_{S/R}\approx0.6$; light: $ER_{S/R}\approx0.1$).

3.4
Photochemical Conversion Processes

In general, photochemical transformation of chiral xenobiotics is assumed to be largely nonenantioselective. This hypothesis forms the basis for discrimination between biotic (enantioselective) and abiotic processes like photodecomposition. Systematic laboratory investigations on photochemical transformation of α-HCH and β-PCCH performed by Hühnerfuss et al. [2], however, showed that caution has to be applied when interpreting results obtained by enantioselective gas chromatography.

As a starting point, a mixture of α-HCH and its transformation product β-PCCH was used which had been obtained in the course of the preparation of (–)-α-HCH by reaction of technical (±)-α-HCH with (+)-brucine [3]. After a first preparative cycle, the ratio (+)-α-HCH/(–)-α-HCH was 0.54, and the ratio β_1-PCCH/β_2-PCCH was 1.21. This mixture was exposed to artificial light, the spectrum of which was almost exactly the same as the spectrum of sunlight (using a

Table 3.23. Enantiomeric ratios of α-HCH and β-PCCH formed by photochemical reactions during a period of 35 days. At the beginning of the experiment the enantiomeric ratios for α-HCH and β-PCCH were 0.54 and 1.21, respectively; from [2]

Time (days)	Enantiomeric ratios	
	$(+)$-/$(-)$-α-HCH	β_1-/β_2-PCCH
0	0.54	1.21
5	0.47	1.02
7	0.46	0.96
9	0.48	0.84
12	0.49	0.69
14	0.46	0.67
16	0.47	0.62
21	0.45	0.47
23	0.49	0.51
26	0.48	0.48
29	0.48	0.45
35	0.44	0.44

SOL 500 lamp, Höhnle, Germany). As shown in Table 3.23, the enantiomeric ratio $(+)$-α-HCH/$(-)$-α-HCH remained constant throughout the experimental period of 35 days, exhibiting an average value of 0.48±0.04. The same value was approached by the ratio β_1-PCCH/β_2-PCCH after a photochemical reaction of about three weeks. This implies that the photochemical transformation of β-PCCH that had been present in the starting mixture was complete after about three weeks; however, additional β-PCCH was continuously formed by photochemical transformation of α-HCH nonenantioselectively. Since the α-HCH present in the mixture exhibited an enantiomeric ratio of about 0.48, the non-enantioselective formation of β-PCCH in this case also gave rise to an enantiomeric ratio of about 0.48.

The groups of Buser and Müller [47] and Parlar and co-workers [22] studied the photoconversion products of cyclodiene insecticides, because of their higher toxicity and stability in the environment in comparison with their original pesticides. Heptachlor, *cis*- and *trans*-chlordane, *cis*- and *trans*-nonachlor, and the technical chlordane were exposed to natural sunlight in the presence of air by Buser and Müller [47]. The photoconversion products from heptachlor and *cis*-chlordane were identified by MS as photoheptachlor and two photo-*cis*-chlordanes with caged and halfcaged structures, respectively. Basically, the three photoproducts are chiral; however, the analysis of photoheptachlor by enantioselective cGC clearly proved that it is formed in enantiomeric ratios of 1:1, consistent with a nonenantioselective transformation of the racemic parent compound heptachlor. In the case of the two photo-*cis*-chlordanes, which are formed in an approximate 2:1 ratio, only two partially resolved peaks were found in the GC chromatogram, which Buser and Müller attributed to the two isomers and not to any pair of enantiomers. As the separation of the enantiomers was unsuccessful in the latter instance, no judgment on the selectivity of this process is possible thus far. Analogous products were not observed from *trans*-chlordane or *cis*- and *trans*-nonachlor.

Parlar and co-workers [22] were the first to report the application of enantio-selective cGC to the enantiomer separation of the chiral photoconversion products of photodieldrin, photoheptachlorepoxide, and photochlordene. In all three cases as well as for photoheptachlor, which was also included in their studies, enantiomeric ratios of 1:1 were verified.

In summary, the photochemical transformation of all chiral xenobiotics thus far investigated by enantioselective cGC was shown to be effected nonenantioselectively, as is expected from photochemical processes. However, enantiomeric excesses of transformation products that may have been formed by preceding enzymatic processes may be modified by photochemical processes, as shown above for α-HCH and its transformation product β-PCCH, provided that enzymatic processes become less important, e.g., due to seasonal variations in microbial activity [2].

3.5
Transformation/Accumulation of Chiral Xenobiotics in Sediments and Soils

3.5.1
Chiral Xenobiotics in Sediments

Time profiles of xenobiotics, for example toxaphenes, in sediment cores often supply insight into time-dependent variations in application and/or transformation of these compounds. As most toxaphene congeners are chiral, Vetter et al. [145, 146] raised the question as to whether the most abundant congeners in sediment cores, the heptachlorobornane B7–1001 ("Hp-Sed") and the hexachlorobornane B6–923 ("Hx-Sed"), are accumulated in racemic composition and/or their formation/retention in sediments is accompanied by a change in the enantiomeric ratios. The sediment core investigated by Vetter et al. was collected from Hanson Lake in the Yukon Territories, Canada, in 1994. From 1959 to 1971, Yukon Fisheries carried out a program to facilitate trout fishing close to major population centres. As part of this plan, in the interval July 11–14, 1963, Hanson Lake was treated with toxaphene (0.006 ppm) to kill off the existing populations of whitefish and northern pike. In an effort to restock the lake with trout, in 1965, 1966 and 1968 rainbow trout eggs were planted in the creek joining the two Hanson Lake basins. Recent surveys, however, showed that the only fish species present in the lake today is northern pike.

Vetter et al. [145, 146] analysed sediment cores dated from 1935 to 1992, using three different chiral stationary phases:
- 10% immobilised permethylated β-cyclodextrin (CP-Chirasil-Dex, Chrompack, Middelburg; abbreviated β-PMCD);
- 25% *tert*-butyldimethylsilylated β-cyclodextrin diluted in 85% dimethyl, 15% diphenyl polysiloxane (BGB Analytik, Adliswil, Switzerland; abbreviated β-BSCD); and

- 35% heptakis(6-*O*-*tert*-butyldimethylsilyl-2,3-di-*O*-methyl)-β-cyclodextrin (TBDMS-CD) diluted in OV-1701.

Although enantioseparation of B6–923 failed on β-PCMD, the enantiomers of B7–1001 were resolved. However, the enantiomer separation of both B7–1001 and B6–923 was obtained on β-BSCD. Interestingly, both compounds eluted in reversed order from β-PMCD and β-BSCD. On β-PMCD, the nonresolved B6–923 enantiomers eluted prior to the resolved enantiomers of B7–1001, while β-BSCD eluted the enantiomers of B6–923 after the enantiomers of B7–1001.

Table 3.24 lists the enantiomeric ratios of these two toxaphene congeners in the respective sediment cores. For B7–1001, Vetter et al. found a significantly enantio-enriched second-eluting enantiomer, resulting in ER values from about 0.7 to 0.8. The values on both chiral selectors mainly used for the analyses of the sediment core extracts agreed well with some deviations in cores #5 and #6. On β-BSCD, B7–1001 was interfered in core #3/4 by an unknown compound. Basically, however, a trend towards lower enantiomeric ratios from the early to the most recent cores can be inferred from the results for this toxaphene congener. In contrast, the hexachlorobornane B6–923 was racemic in all samples. This is particularly remarkable, as Fingerling et al. [147] recently obtained B6–923 from higher chlorinated bornanes by reductive dechlorination in soil under anaerobic conditions, and it turned out that they only found one enantiomer under such experimental conditions. This may point to the conclusion that the type of microorganisms in the respective sediments or soils may be crucial for the enantioselective transformation of toxaphenes.

Another study on chiral toxaphene congeners in sediment was carried out by Rappe et al. in the Baltic Sea [148]. The samples were collected close to the coast (4 km from a pulp mill) and in the open sea approximately 150 km off the coast

Table 3.24. Enantiomeric ratios of the toxaphene congeners B7–1001 and B7–923 in sediment cores from the Canadian Hanson Lake as determined by application of two different chiral selectors, β-PMCD and β-BSCD. The results for different ECNI-SIM masses are given; from [145, 146]

| Toxaphene congener | B7–1001 | | | | B6–923 | |
| | β-PMCD | | β-BSCD | | β-BSCD | |
core # (date)	ER m/z 343	ER m/z 345	ER m/z 343	ER m/z 345	ER m/z 307	ER m/z 309
#2 (1992)	0.69	0.74	0.71	0.71	0.96	1.02
#3/4 (1984/7)	0.70	0.70	–	–	0.99	1.01
#5 (1979)	0.68	0.70	0.77	0.76	0.98	1.00
#6 (1973)	0.72	0.71	0.78	0.78	0.98	0.98
#7 (1968)	0.75	0.75	0.81	0.80	0.98	1.00
#8 (1964)	0.81	0.81	0.82	0.82	1.06	1.03
#9 (1959)	0.79	0.79	0.82	0.84	0.97	0.97
#10 (1954)	0.81	0.81	0.81	0.82	1.01	1.02
#11 (1946)	0.79	0.79	0.80	0.78	1.00	0.99
#12 (1935)	0.77	0.77	<<1	<<1	0.97	1.02

in the years 1985–1986. It should be noted that during the sampling period free chlorine was still used for bleaching. Hexa- to nonachlorobornanes were detected in both the "coastal" and the "open-sea" samples at total concentrations estimated to be in the low ng/g range (dry-weight). The isomer pattern of these two sediment samples indicated significant alteration compared with the technical toxaphene mixture.

In the "open-sea" sample a single major hexachlorobornane was detected. Based on retention time, this compound was identified as "Hx-Sed". Two major heptachlorobornane congeners were identified: "Hp-Sed" (sometimes referred to as TC1) and another congener of unknown structure TC2. The "coastal" sample was both similar and dissimilar to the "open-sea" sample depending on the congener group considered. The same heptachlorobornane congeners were present, but "Hp-Sed" was higher in the "coastal" sample. "Hx-Sed" was present in the sediments from both locations.

The enantioselective analyses were carried out with a stationary phase consisting of a mixture of OV-1701 and β-BSCD. While "Hx-Sed" was only marginally resolved, as indicated by an increased peak width, the two major heptachloro congeners, "Hp-Sed" and TC2, were clearly resolved. Rappe et al. conclude *that the chromatograms indicate enantiomeric ratios for the heptachloro congeners in the sediment somewhat differing from an exact 1:1 ratio.*

Benická et al. [149] analysed selected PCB atropisomers (PCB95, PCB91, PCB84) in extracts from a river sediment sample by multidimensional gas chromatography in combination with a Valco two-way switching valve and two high-resolution fused-silica columns (achiral phase: CP-Sil 8; chiral phase: CP Chirasil-Dex CB). For comparison, the respective standard PCBs, as well as a technical PCB mixture (1:1 mixture of Arochlor1242 and Arochlor1260), were included in the study.

In the technical mixture, the enantiomer peaks of PCB95 were separated with a resolution factor of 0.98, i.e., no clear baseline separation was achieved, and, as a consequence, the area ratios of the two peaks slightly differed from 1.0, which would have been the expected value for the racemic congener. The quotient of the peak areas supplied an ER value of 0.972, after a deconvolution procedure an ER value of 0.984 was calculated (for the pure PCB95 standard after deconvolution ER is 0.99). In the sediment sample extract, separation of PCB95 enantiomers was attained without apparent distortion of the peaks. However, Benická et al. point out that the area ratio for the PCB atropisomers and the shape of the second peak were highly dependent on even a slight shift of the heart-cut interval. This indicated the presence of an impurity, a conclusion that was also inferred from deconvolution of the two chromatograms. An interfering peak hidden under the peak of the second enantiomer caused a difference between PCB95 enantiomeric ratios as calculated from the original chromatogram (ER 0.605) and after the deconvolution procedure (ER 0.696). These results clearly indicate a depletion of the first-eluting PCB95 enantiomer in the river sediment.

Analysis of the PCB91 atropisomers turned out to be less complicated, because low interference was encountered on the first achiral column. The resolution factor for the PCB91 enantiomers was 0.64. For the PCB91 standard an enantiomeric ratio of 0.930 was determined, while in the sediment sample extract an ER value of 0.936 was found. This result implies that no significant enzymatic transformation of this PCB congener occurred in the river sediment. Analysis of the PCB84 atropisomers was more complicated, because an overlap of the PCB congeners 92 and 89 can occur, and also additional PCBs may interfere with the PCB84 peaks. By deconvolution of the original chromatogram, five peaks were found under the real chromatogram envelope. Therefore, additional investigations are due to solve this separation problem.

3.5.2
Chiral Xenobiotics in Soils and Ambient Air

In this section, emphasis will be placed on investigations that focus on transformation and accumulation of chiral pollutants in soils. Furthermore, the aspect of transport processes between soil and air will be discussed. With regard to the problem related to "chirality and crop protection", the reader should refer to the comprehensive reviews published by Ariëns et al. [150], Ramos Tombo and Belluš [151], and Buser and Francotte [152].

Organochlorine (OC) pesticides were used heavily on farmlands in the United States and Canada during the 1960s and 1970s. As OC pesticides and their metabolites are highly persistent, residues remained in many soils. For example, chlordane and heptachlor have not been used in agriculture since 1983, but before that time 65% of chlordane applications were for crops, home lawns, turf and ornamentals (see refs in [153]). From 1983 until their ban in 1988, chlordane and heptachlor were used exclusively for structural termite control, resulting in greatly elevated levels in home air. Based on this background, an interdisciplinary US/Canadian/ Swedish working group started an investigation which focused on the question as to whether or not sources of chlordane and heptachlor to ambient air still include volatilisation from agricultural soils and/or emissions from house foundations or are largely caused by long-range transport from countries still using these OC pesticides such as Mexico and other Latin American countries [153–156]. Furthermore, the authors addressed the problem as to whether heptachlor *exo*-epoxide in the air arises mainly from photolysis of heptachlor or by volatilisation of this compound as produced by metabolism of heptachlor in soils.

In the course of this study enantioselective gas chromatography played a major role, because it was conjectured that the chiral pesticides chlordane and heptachlor, which were and still are being applied as racemic mixtures, are being emitted from houses as well as from actual long-distant application sites as racemates. Once in the air, photolysis may yield oxychlordane, photoheptachlor, heptachlorepoxide and other oxidation products, which in all cases will form racemates. By contrast, Aigner et al. showed that selective enzymatic degradation of chlordane enantiomers by microorganisms did occur in soils from the Midwestern United States resulting in nonracemic signatures of the chiral pesticides in soils [157]. The same is assumed to be valid for heptachlor. This result formed the basis for the application of nonracemic enantiomeric signatures to distinguish between releases of chlordane and heptachlor from agricultural soils vs. termiticide emissions.

Soil and air samples were collected in four states in the U.S. Cornbelt region: Ohio, Pennsylvania, Indiana and Illinois, at different times from 1995–1997. At 38 farms, eight soil cores ($\approx$15 cm depth) were taken and pooled to obtain a representative sample for each field [156, 157]. Three types of air samples were collected: ambient, above soil and indoor, using polyurethane foam (PUF) traps (for experimental details, see [156]). Ambient and indoor samples were also collected around Muscle Shoals, Alabama, and ambient and indoor air in Columbia, South Carolina [154].

Chlordane residues (sum of *cis*- and *trans*-chlordane) in Cornbelt soils were, on average, ten times higher than levels found in soils from Alabama (geometric means: Cornbelt 1.4 ng/g; Alabama 0.17 ng/g [154, 156]). Average concentrations of chlordanes in ambient air in the Cornbelt (0.041 ng/m^3, n=4) were similar to annual means for chlordane at Sturgeon Point on Lake Erie (0.034 ng/m^3; refs in [156]). Ambient levels in two southern states were somewhat higher at 0.109 ng/m^3 for rural Alabama and 0.295 ng/m^3 in the city of Columbia, SC [158]. Average indoor air concentrations in the Cornbelt states (16.7 ng/m^3, n= 20) were similar to average levels in homes in Columbia (10.1 ng/m^3, n=3) and rural Alabama (34.2 ng/m^3, n=5).

Average enantiomeric ratios for *trans*- and *cis*-chlordane, heptachlor and heptachlor *exo*-epoxide in air and soils from the Cornbelt region, South Carolina and Alabama are summarised in Table 3.25. Soils in the Cornbelt region show an enantiomeric difference for both *trans*- and *cis*-chlordane: in the case of *trans*-chlordane ER values less than 1.00 indicate preferential transformation of the (+)-enantiomer, while for *cis*-chlordane ER values greater than 1.00 imply a preferential transformation of the (–)-enantiomer. A comparison of enantiomeric ratios for Cornbelt soils and air-above-soil samples was carried out for six fields. The enantiomeric signatures of pesticides in the air above the soil followed the same pattern as in the soil, both in direction of transformation and in relative magnitude. This preservation of ER profile on volatilisation is in line with the results obtained for heptachlor, heptachlor *exo*-epoxide and α-HCH [155, 156, 159]. Chlordane ER values in Alabama soils were also nonracemic, but the difference was smaller than in the Cornbelt soils.

Enantiomeric ratios for ambient air samples taken at rural, nonagricultural locations in the Cornbelt show the same general trend (although less pronounced) as was seen in soil and air-above-soil samples. Ambient air ER values in rural locations show transformation of the (–)-enantiomer. A comparison of the enantiomeric ratios for Alabama and in Columbia, SC, were closer to racemic. In contrast, air samples over the Great Lakes show a depletion of (+)-*trans*-chlordane and a slight depletion of (–)-*cis*-chlordane enantiomers, similar to the ERs observed for Cornbelt soils. ER values of chlordanes in air from homes in the Cornbelt states were all very close to racemic. Racemic chlordane

Table 3.25. Average enantiomeric ratios (ER≡(+)-/(–)-enantiomer; ±standard deviation) for *trans*- and *cis*-chlordane, heptachlor and heptachlor *exo*-epoxide in air and soils from the Cornbelt region, South Carolina and Alabama; from [155, 156]

Location	*trans*-Chlordane ER	*cis*-Chlordane ER	Heptachlor ER	Heptachlor *exo*-epoxide Conc. (pg/m^3) ER	
AIR					
Above-soil, Cornbelt (n=6)	0.74 (0.10)	1.11 (0.08)			
Ambient, Cornbelt (n=3)	0.93 (0.03)	1.04 (0.04)			
Indoor, Cornbelt (n=15)	0.99 (0.01)	0.98 (0.03)			
Ambient, Alabama (n=20)	0.98 (0.03)	1.01 (0.04)			
Ambient, SC (n=7)	1.00 (0.01)	1.02 (0.01)			
Indoor, AL & SC (n=8)	0.98 (0.01)	1.00 (0.01)			
Lake Ontario (n=15)	0.92 (0.02)	1.03 (0.02)			
Ambient, Columbia SC (n=9)			0.99/1.02 (n=2)	–	1.51 (0.09)
Ambient, Muscle shoals, AL (n=6)			0.98 (0.02) (n=4)	20±19 (n=7)	1.70 (0.18)
Ambient, Point Petre, Lake Ontario (n=8)			1.01 (0.05) (n=6)	7.3±3.9 (n=7)	1.86 (0.05)
Ambient, Lake Superior (n=6)			–	19±12	2.02 (0.10)
SOIL					
Cornbelt region (n=38)	0.70 (0.12)	1.21 (0.14)			
Alabama (n=9)	0.91 (0.05)	1.11 (0.10)			

was also found in indoor air for several homes in Columbia, SC and Alabama [154].

The data set available until 1998 allowed the following conclusions with regard to potential sources for chlordanes in the air [156]: High concentrations of chlordanes (compared with average Great Lakes values) in indoor and ambient air from Columbia, South Carolina and intermediate levels near the town of Muscle Shoals, Alabama, go along with racemic mixtures in the air in both locations, while nonracemic chlordanes were reported in southern soils. These findings suggest that termiticide-treated houses, rather than soil emissions, are the main source of chlordane to southern U.S. air. Chlordane concentrations in home air from the Cornbelt region are much greater than average ambient air concentrations from the Cornbelt region or the Great Lakes. However, enantiomeric ratios of chlordane in ambient air from the Cornbelt and the Great Lakes lie between the ER values in air-above-soil (nonracemic) and home air (racemic). Lower concentrations coupled with nonracemic ER values imply a greater import of agricultural sources in these regions, i.e., a mixture of termiticide-treated home air and emissions from regional soils can be assumed.

The interpretation of the results for air/soil exchange of heptachlor included its main transformation product heptachlor *exo*-epoxide, also a chiral compound and thus accessible to enantioselective gas chromatography [155]. Concentrations of heptachlor *exo*-epoxide in ambient air in Alabama ranged from 4–9 pg/m^3 in January/February to 29–51 pg/m^3 in May/June, similar to the seasonality reported in southern Ontario (refs. in [155]). This is consistent with temperature-driven volatilisation from soil. The enantiomeric composition of this compound in air samples from all locations was distinctly nonracemic (Table 3.25). Average ER values [(+)–/(–)-enantiomer] ranged from 1.51 in Columbia to 2.02 over Lake Superior. The enantiomeric ratios were remarkably consistent in each location, exhibiting relative standard deviations of only 2.7–10.6% and showing no seasonal dependence.

Heptachlor was quantified only in the air samples from Muscle Shoals and, unlike heptachlor *exo*-epoxide, showed no seasonality, ranging from 27–45 pg/m^3 in January/February to 31–49 pg/m^3 in May/June. The enantiomers of heptachlor were only partially resolved on the BGB-172 column used by Bidleman et al. [155], although well enough to determine the enantiomeric ratios in air samples. The average ER values of heptachlor were 1.01±0.05 (n=6) at Point Petre and 0.98±0.02 (n=4) in Muscle Shoals, and they were not significantly different from the standard. Enantiomeric ratios of heptachlor in two air samples from Columbia were 0.99 and 1.02. These ER values are consistent with a study in which racemic heptachlor (ER=1.03) was found in Norwegian air [47].

These results suggest that heptachlor *exo*-epoxide in ambient air does not mainly arise from heptachlor photolysis, although some racemic portion may originate from this source. Release of heptachlor *exo*-epoxide from soils is a more plausible explanation. ER values of this heptachlor metabolite averaged 1.30±0.08 in soils of four British Columbia vegetable farms [160], 2.87±1.52 (n=14) in agricultural soils from the Cornbelt states of Ohio, Illinois, and Indiana [157], and 2.71–3.19 in an agricultural and cemetery soil from Alabama [154]. The heptachlor *exo*-epoxide in air samples collected 5–140 cm above the soil at the British Columbia farm showed the same ER as in the soil, suggesting soil to air transfer. Thus, this metabolite is likely produced in soil by enantioselective epoxidation of heptachlor and subsequently volatilised. In the Great Lakes region, it is also possible that volatilisation from water contributes some heptachlor *exo*-epoxide to the atmosphere. However, nonracemic heptachlor *exo*-epoxide was also found in Columbia, SC, and Muscle Shoals, AL, well away from large bodies of water.

In this context, Bidleman et al. raised the question as to how the presence of racemic heptachlor in ambient air can be explained [155]. If (+)-heptachlor in soil undergoes preferential epoxidation to (+)-heptachlor *exo*-epoxide, the residual heptachlor should be nonracemic and have an ER <1. However, this was not observed in British Columbian soils, where the ER of heptachlor was 1.08 despite selective formation of (+)-heptachlor *exo*-epoxide (ER 1.30). Heptachlor transformation in soil proceeds by at least two routes. Following its incorporation into the soil of an experimental field in Ohio, heptachlor was dissipated with a half-life of 0.91 year (see refs in [155]). Two degradation products were identified: 1-hydroxychlordene and the more persistent metabolite heptachlor *exo*-epoxide. After 4.5 years, heptachlor was not detectable and heptachlor *exo*-epoxide accounted for about 20% of the original heptachlor applied. Heptachlor is more volatile than its epoxide and would be expected to dissipate more quickly by evaporation. Perhaps only a small proportion of the heptachlor in soil is converted to the epoxide, and the rest is volatilised or degraded by processes that are not enantioselective or have different preference for the heptachlor enantiomers. Another possibility discussed by Bidleman et al. [155] is that racemic heptachlor is released from buildings that were protected against termites with heptachlor or technical chlordane. Although the enantiomeric composition of heptachlor in home air has not been determined, the *cis*- and *trans*-isomers of chlordane in the air of eight homes in Columbia and Muscle Shoals were racemic, suggesting that the heptachlor in this air would also be racemic. Atmospheric transport of recently applied heptachlor from outside the United States or Canada could also contribute racemic heptachlor to ambient air.

Lewis et al. [164] showed in field and laboratory experiments that environmental changes in soils can alter the preferential transformation of chiral environmental pollutants. The authors intended to assess the persistence of different pollutant enantiomers across a wide latitudinal transect, and the possible influence of large-scale environmental changes on that persistence. They investigated enantioselective transformation in soils collected from three areas after amending them with ruelene [i.e., (*R,S*)-4-*tert*-butyl-2-chlorophenylmethyl-*N*-methyl phosphoramidate], dichlorprop [DCPP, i.e., (*R,S*)-2-(2,4-dichlorophenoxy)propanoic acid], and methyl dichlorprop (see Fig. 3.33). Only the (+)-enantiomers of dichlorprop and methyl dichlorprop are herbicidal, while both enantiomers of ruelene are herbicidal, though, in the latter case, the (+)-enantiomer is four times more toxic. The experimental sites included an upland plateau in Risdalsheia, Norway, about 20 km inland of the North Sea, an 80-year-old mixed deciduous forest (Harvard Forest) in the north-eastern U.S., and an area of the Fazenda Nova Vida about 250 km south of the city of Porto Velho in Rondonia, Brazil.

Regarding the Brazilian soil samples, emphasis was placed on the question as to whether or not tropical deforestation (i.e., conversion to pastures) may have caused an effect on the enantioselectivity of transformation or on demethylation rates of the racemate of methyl dichlorprop. Soil samples were collected from a forest in Rondonia and from six deforested areas (pastures) nearby, ranging be-

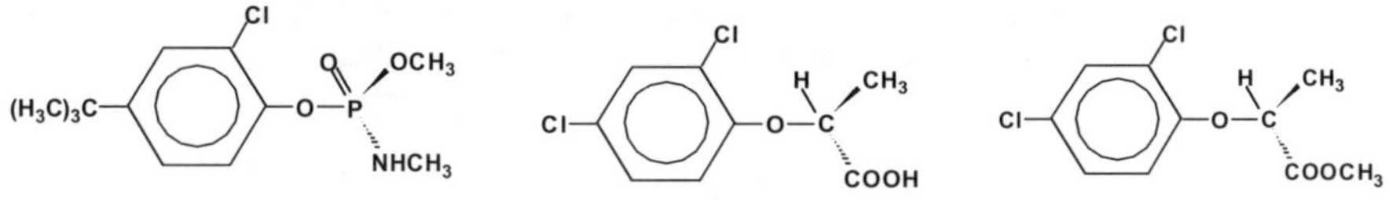

Fig. 3.33. Molecular structures of ruelene [i.e., (*S*)-4-*tert*-butyl-2-chlorophenylmethyl-*N*-methyl phosphoramidate], dichlorprop [i.e., DCPP; (*R*)-2-(2,4-dichlorophenoxy)propanoic acid], and methyl dichlorprop

Table 3.26. Effect of tropical deforestation (i.e., conversion to pastures) on enantioselectivity of transformation or demethylation rates of the racemate of methyl dichlorprop as well as on enantioselectivity of dichlorprop (DCPP) transformation in Brazilian soils; modified from [164]

Sample	n	$t_{1/2}$ (days)	C.V. (%)	(+)- selective (%)	(−)- selective (%)	(+/−)- nonselective (%)
Methyl dichlorprop						
Forest	52	0.77	62	7.7	1.9	90
Pasture	75	0.63	49	11	1.7	87
Dichlorprop						
Forest	16	11	36	13	38	50
Pasture	37	16	57	6.7	93	0.0

tween 9 and 47 years in age. One gram of each soil sample was amended with the respective herbicide (see Fig. 3.33) and incubated overnight for methyl dichlorprop and over about 6 months for dichlorprop. Lewis et al. used a strange way of summarising their results, which at first glance may cause some difficulties to the reader (Tables 3.26 and 3.27): selectivity for different enantiomers, (+)- or (−)-enantiomer, was documented by the authors *as percentages of soil samples in which differences in the amounts transformed of each enantiomer were ≥15%.* Nonselectivity (+/−) was inferred from the data if the concentration differences between the (+)- and (−)-enantiomer were <15%. Furthermore, half-lives ($t_{1/2}$) of herbicide racemates were calculated ± a coefficient of variation (C.V.) among a number (*n*) of replicate soil samples.

The results summarised in Table 3.26 show that tropical deforestation had little or no effect on the enantioselective transformation or on the demethylation rates of racemic methyl dichlorprop. Obviously, demethylation was the first and most rapid step in the transformation process, which quantitatively produced dichlorprop as the reaction product. Soil warming in North America and Norway had no measurable effect on demethylation rates or enantioselective transformation characteristics when field plots were heated 5 K above ambient temperatures for about 7 years in North America and about 4 years in Norway. Transformation of dichlorprop began after a 6-month lag period and proceeded 20 times more slowly (half-life of about 14 days for dichlorprop compared with

Table 3.27. Effect of organic nutrient enrichment (beef extract and peptone) on enantioselectivity of transformation for racemic methyl dichlorprop; modified from [164]

Sample	n	(+)-selective (%)	(–)-selective (%)	(+/–)-nonselective (%)	Net preference (%)
Rondonia, Brazil					
Soil	127	9.0	2.0	88	7.0 (+)
Nutrient enrichment	340	15	50	35	35 (–)
Harvard Forest, MA USA					
Soil	29	45	48	7.0	3.0 (–)
Nutrient enrichment	207	3.0	80	17	77 (–)
Risdalsheia, Norway					
Soil	16	0.0	88	13	88 (–)
Nutrient enrichment	80	15	74	11	59 (–)

about 0.7 days for methyl dichlorprop). Addition of inorganic fertilisers to forest and pasture plots in Brazil had no measurable effect on enantioselectivity for methyl dichlorprop (tested at 2 weeks and 5 months after fertilisation), or to plots in North America, fertilised since May 1988.

Laboratory enrichment experiments (addition of beef extract and peptone) caused a shift in the preferential transformation characteristics: soil samples which showed a preferential transformation of the (+)-enantiomer of methyl dichlorprop after addition of the organic nutrient shifted to a preferential transformation of the inactive (–)-enantiomer in samples from North America and Brazil (Table 3.27). Soils from Norway already in the original sample mainly removed the (–)-enantiomer. Lewis et al. [164] conjecture that microorganisms preferring the (+)-enantiomer may have increased in activity after organic nutrient amendments (Table 3.27), but most samples (74%) still preferred the (–)-enantiomer. The authors pointed out that preferential removal of (–)-methyl dichlorprop would render residues more phytotoxic than the same concentrations of the racemate, as long as demethylation is a prerequisite step and the (+)-enantiomer is not preferentially removed after nutrient enrichment.

Ruelene, like dichlorprop, was transformed by microorganisms only after a time lag of about 6 months. The half-life, once transformation began, was about 40 days in Brazilian soils and 20 days in Norwegian soils. Furthermore, Lewis et al. [164] concluded that the transformation rates of racemic ruelene were unaffected by deforestation or soil warming. Enantioselectivity of the transformation process was, however, influenced in both cases. In Brazil, forest soils, originally removing preferentially the less toxic (–)-enantiomer [i.e., 22% (+), 67% (–), 11% (+/–); *n*=17], after deforestation shifted to transforming exclusively the (+)-ruelene. Soil warming in Norway, on the other hand, caused soils to shift from all samples preferentially removing the (+)-ruelene to some (22%) removing the less toxic (–)-enantiomer (*n*=13).

Delayed microbial transformation of chemical residues is commonly observed and is often assumed to represent the amount of time required for low concentrations of microorganisms responsible for the transformation process to attain significant numbers after using the chemicals as substrates or energy sources. In addition to this general aspect, Lewis et al. [164] gained deepened insight into the dependence of the enantioselective transformation processes on the microorganisms actually present in the respective soils. The diversity and biogeography of bacteria capable of demethylating methyl dichlorprop were investigated by comparing small-subunit (16 S) ribosomal RNA (rRNA) genes of 50 soil bacterial isolates using amplified ribosomal DNA restriction analysis (ARDRA). Furthermore, ARDRA has proved useful in taxonomic assessments of bacterial culture collections and 16 S rDNA clone libraries at genus and species levels (for details of these methods see refs in [164]). In general, isolates from Norway and North America formed three large clusters and appeared to be more closely related than to isolates from Brazil. Conversely, Brazilian isolates formed smaller clusters that were more distantly related to one another and to Norwegian and North American clusters. It appears to be plausible that these relationships among bacterial isolates reflect ecological differences between these geographical regions. ARDRA patterns that indicated taxonomic identity turned out to exhibit gene-sequence identity (except for two isolates from Norway).

A comprehensive interpretation of the results obtained from enantioselective analysis and from the ARDRA pattern supplied the notable result that isolates with identical ARDRA patterns exhibited the same enantioselectivity phenotype regardless of their geographical origins, except for the two isolates from Norway mentioned above. For more details on the enantioselectivity of the different bacterial isolates the reader is referred to the original paper [164]. In conclusion, Lewis et al. suggest that microbial enantioselectivity with environmental pollutants is controlled by the activation of metabolically quiescent microbial populations or the induction of enantiomer-specific enzymes, as was the case with amino acids, and that this selectivity follows lines of genetic similarity where different groups of related microbes are activated by different kinds of environmental changes.

3.6
Air/Water Gas Exchange and Atmospheric Long-Range Transport

3.6.1
Air/Water Gas Exchange Studies in Lakes

Air/water gas transfer involves both absorption and volatilisation; the direction of net exchange is usually determined by the difference between the concentrations of the gaseous and dissolved compound in air and surface water. The magnitude of the exchange depends on this difference and the mass transfer coefficient that involves the temperature-dependent Henry's law constant. This latter aspect may be oversimplified because of uncertainties in the Henry's law constant and the mass transfer coefficient. Furthermore, on an undulating water surface the presence of capillary waves, wave breaking processes as well as slick-induced modifications of these processes may strongly influence air/water exchange processes [161]. In addition, such methods are subject to analytical errors in measuring the concentrations of the compound(s) in air and water. Together, these uncertainties are substantial and may sum up to several hundred percent [13]. Enantioselective gas chromatography of chiral pollutants helped to circumvent the latter analytical errors and provided new insights into the environmental transport of toxic compounds as well as information pertaining to the environmental transformation of these compounds. As an advantage, the relative peak areas of enantiomers or enantiomeric ratios can be determined with high precision (relative standard deviations generally better than ±4%) by enantioselective cGC, whereas the precision in measuring concentrations of trace organic compounds is typically 15–20% [13], and computing the difference of two concentrations thus yields an even greater error. Therefore, enantioselective cGC can be used to examine environmental processes difficult to resolve by comparing concentration data alone.

One chiral pollutant that is prevalent in the environment and partitions readily between the atmosphere and surface waters of lakes and the oceans because of its relatively low Henry's law constant is α-HCH. As a result of gas phase absorption, it is estimated that about 20% of the global environment burden of HCHs is held in the upper 200 m of the world's oceans [13]. Therefore, very extensive investigations have been carried out by Bidleman and co-workers including enantioselective gas chromatographic analyses of α-HCH [11–16, 58, 59, 153]. For example, air/water interaction studies in Lake Ontario water, rain and air samples supplied a comprehensive survey [13, 14, 153]. The α-HCH rain samples collected at Point Petre were racemic, the ER values ranging from 0.98 to 1.02 ($n=7$), while concentration data ranged from 870 to 3030 pg/L. Water samples taken from Western, Central, and Eastern Basins of Lake Ontario were similar and did not differ between April and October 1993. An average ER of 0.86 ± 0.02 ($n=16$) was obtained for surface samples. ERs of α-HCH in water taken from the hypolimnion averaged 0.85 ± 0.02 ($n=29$) and were not significantly different from surface water ERs. Higher enantiomeric ratios were observed in water samples from the Niagra River (0.91 ± 0.02). The enantiomeric compositions of α-HCH in air samples taken on board a ship during cruises across Lake Ontario at 10 m above the lake vary seasonally with near racemic values in the spring and fall and values as low as 0.91 in mid-summer. A simple air/water gas transfer model demonstrated that enantiomeric ratios <1 in air were derived from equilibration of the air with the water during transport of the air mass over the lake. Based on the experimental ERs, Ridal et al. [13] calculated that as much as 60% of the α-HCH in air above Lake Ontario was derived from the lake itself. These results strongly support the need for over-water air measurements to provide better estimates of air/water gas transfer fluxes of persistent organic pollutants, but they also suggest the application of enantioselective gas chromatography as a valuable tool, if chiral pollutants are being included in the course of air/water interaction studies.

As an extension of the Integrated Atmospheric Deposition Network (IADN) project, which has been monitoring deposition of organochlorine compounds to the Great Lakes, Ulrich and Hites determined the spatial trends of chlordane-related compounds in air near the Great Lakes [162]. From August 1994 through September 1995, 48 air samples were taken near Lake Erie, five air samples were taken near Lake Michigan, and six air samples were taken near Lake Superior. While there were only slight differences in the enantiomeric ratios between the various sites, there were considerable differences between the compounds [ER defined as the amount of the (+)-enantiomer divided by the amount of the (–)-enantiomer]. The overall ER for *cis*-chlordane was 1.05±0.02, which is close to racemic. The overall ER for *trans*-chlordane was 0.88±0.02, which is significantly different from racemic and from the respective *cis*-chlordane value. This discrepancy suggests that *trans*- and *cis*-chlordane are metabolised differently in the environment. The overall ER of heptachlor *exo*-epoxide was 1.99±0.04; this large deviation from racemic indicates that this compound largely is an enzymatic transformation product of heptachlor. Wiberg et al. [154] found nonracemic values for the same three compounds in Great Lake air and nonracemic values for heptachlor *exo*-epoxide in southern U.S. air, while Buser and Müller [47] and Wiberg et al. [154] reported ER values near 1.0 for heptachlor, *cis*- and *trans*-chlordane in Norwegian air and in southern U.S. air, respectively.

3.6.2
Air/Sea Gas Exchange Studies

A bridge between air/water exchange studies performed on a lake and on the open sea is being formed by an investigation carried out by Falconer et al. both on Amituk Lake, Cornwallis Island (N.W.T., Canada) and off the coast of Cornwallis Island in Resolute Bay [12]. Air and water samples were taken at Cornwallis Island (74°N, 95°W) in August–September, 1992. Water samples were collected from Resolute Bay and Resolute Passage by small boat or helicopter. Several samples were also taken from Amituk Lake, a small stream-fed lake on the east side of Cornwallis Island. For further experimental details, refer to [12].

In air samples from Resolute Bay, where chances for metabolism were low, an ER of 1.00±0.04 was found for α-HCH, which clearly indicates a racemic mixture. In the seawater samples, the (+)-α-HCH enantiomer was depleted resulting in an average ER of 0.93±0.06. Enantiomeric ratios of Amituk Lake samples differed greatly depending on location and date. Water samples from tributaries of the lake, which are fed by snowmelt but also drain water from other small lakes, exhibited ER values from 0.88–0.99 in late June to 0.65–0.86 in mid-July. The lake water at about 20 m depth showed an average ER of 0.77. On June 30th and July 4th, the outflow from the well-mixed lake had ERs in the range 0.74–0.77. On July 7th and 9th, the outflow value rose to 0.85–0.86, corresponding to peak snowmelt and cold, less dense water flowing across the surface. On August 15th and 16th, the outflow ER returned to 0.74–0.77, as contributions from tributaries decreased. These results suggest that microbial transformation of α-HCH takes place in Arctic lakes and near-shore marine waters. Furthermore, these results encouraged Bidleman and co-workers to expand their enantioselective analyses of air and seawater samples for deepened studies of air/seawater gas exchange studies using similar theoretical approaches to those discussed in Section 3.6.1.

In the summer of 1993, the Russian ship Okean completed a 50-day cruise (BERPAC-93) of the Bering and Chukchi Seas, where water and air sampling for HCHs was included (for details see [15]). In July–September, 1994, another series of samples was collected from the Canadian research vessel Louis S. St. Laurent on a transect of the Arctic Ocean and adjacent waters from Victoria, British Columbia, to Halifax, Nova Scotia, across the north pole. Water and air sampling media were processed for analysis by previously described methods. Extract volumes were brought to 1 mL (quantitative analysis) or 100 µL (enantiomer analysis) by blowdown with nitrogen into *iso*octane (for refs see [15]). Results of the enantiomeric ratios of α-HCH in the water samples of the latter cruise have already been presented in Section 3.1.2. Samples from the Canada Basin and Greenland Sea were depleted in (+)-α-HCH, resulting in ER values of <1.00. An opposite enantiomer degradation pattern was found in the Bering and Chukchi Seas, where the ERs in surface water were generally >1.00, indicating selective breakdown of (–)-α-HCH.

The α-HCH in air samples taken in 1993 and 1994 within 40 m of the sea surface was nonracemic and followed the same order of degradation as in surface water (Fig. 3.34). Air over the Bering Sea and southern Chukchi Sea was depleted in (–)-α-HCH (ER >1.00) but, as with the water, the selectivity reversed at higher latitudes. ERs in air over portions of the Arctic Ocean and the northern Atlantic Ocean were <1.00, from depletion of (+)-α-HCH. The similarities in air and wa-

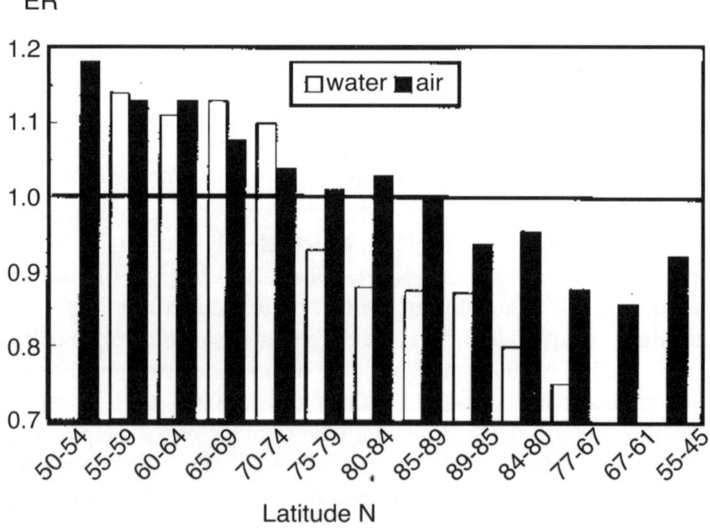

Fig. 3.34. Enantiomeric ratios (ERs) of α-HCH in air and water at different latitudes [ER= (+)-/(–)-α-HCH]; from [15]

ter enantiomeric profiles shown in Fig. 3.34 for the Bering and Greenland Seas suggest that sea-to-air gas exchange is an important source of α-HCH to the marine boundary layer.

Jantunen and Bidleman stress that air/water gas exchange is a "two-way street". At equilibrium the net flux is zero, but volatilisation and deposition still occur at equal rates. This concept has been applied to the exchange of enantiomers (which have the same Henry's law constant!) when the α-HCH in the bulk air is racemic and the α-HCH in surface water is nonracemic [13, 15]. The concentration of α-HCH is about 10^4 greater than in air. Because of this buffering capacity the ER of α-HCH in air will tend toward the seawater value regardless of whether air/water equilibrium is approached from the deposition or volatilisation side. Thus a nonracemic ER value for air does not in itself indicate net volatilisation of α-HCH, although this is implied from the fugacity ratios [15]. The proportion of α-HCH in the atmosphere that has volatilised from the ocean can be estimated from the ER values in boundary-layer air (ER_{bl}) and surface water (ER_{sw}) assuming that the ER of volatilised α-HCH is equal to ER_{sw} ($\neq 1.00$) and the α-HCH in bulk air is racemic (ER$=1.00$). The fraction (f) of α-HCH in the air column arising from outgassing is [178]

$$f = \frac{(ER_{bl}-1)(ER_{sw}+1)}{(ER_{sw}-1)(ER_{bl}+1)}$$

Thus for $ER_{bl}=0.90$ and $ER_{sw}=0.80$, $f \approx 0.5$.

Furthermore, it is worth noting that, according to Fig. 3.34, α-HCH in air over the Canada Basin north of 75°N is racemic even when the water is not. This is possibly due to inhibition of sea-to-air gas exchange by ice cover, although the fugacity ratios predict that the water is oversaturated [15].

In 1996, Harner et al. [59] carried out air/water gas exchange studies of hexachlorocyclohexanes and the enantiomers of α-HCH in the Barents Sea and eastern Arctic Ocean using a similar experimental and theoretical approach to that of Jantunen and Bidleman [15]. Air and water samples were collected aboard the Swedish icebreaker Oden during July to September, 1996; water samples were taken with a submersible pump lowered about 3 m below the water line, while air samples of about 600 m^3 were drawn over 24 h through a glass fibre filter followed by a polyurethane foam (PUF) trap.

The enantiomeric ratios of α-HCH in surface water ranged from 0.72–0.94 and averaged 0.87 ± 0.06 ($n=21$), indicating selective transformation of (+)-α-HCH. Mean ERs in four different zones that can largely be characterised by different latitudes (see also Fig. 1 in [59]) were as follows: latitudes 73–79°N: ER$=0.91 \pm 0.01$; latitudes 80–87°N: ER$=0.83 \pm 0.03$; latitudes 85–88°N: ER$=0.89 \pm 0.04$; latitudes 82–87°N: ER$=0.83 \pm 0.09$. Enantioselective breakdown of (+)-α-HCH was greater in subsurface water, with ERs of about 0.2 to 0.3 at 250–1000 m, agreeing with results from the western Arctic Ocean [16]. The range of enantiomeric ratios in air samples was 0.87–1.00, with a mean of 0.95 ± 0.03 ($n=16$). This result suggests that the air sampled from the ship contained a mixture of nonracemic α-HCH from volatilisation and racemic α-HCH transported from continental regions.

Organochlorine contamination in the Northern Hemisphere has been studied much more extensively than in the Southern Hemisphere, but organochlorine residues, including hexachlorocyclohexanes, have been found in Arctic air, water and biota (see refs in [58]). The global fractionation hypothesis predicts that chemicals released in the temperate and tropical zones will be transported to the colder boreal and polar zones, where they will have the tendency to residue for long periods of time. Therefore, it appeared to be of high interest as to whether indications of enzymatic transformation of α-HCH can be found in Antarctic regions. Between December 1997 and February 1998, air and water samples were collected in the South Atlantic and southern Ocean to determine the air/water gas exchange of α- and γ-HCH and the enantiomeric ratios of α-HCH. The cruise track of the S A Agulhas was as follows: the research vessel left Cape Town, South Africa, on November 30th, 1997, and proceeded south along 6°E reaching the South African National Antarctic Expedition (SANAE) Base (70°S, 3°E), then returning to Cape Town on February 7th, 1998.

Water samples of 20–80 L were collected by a submersible pump beneath the ship. Air samples were taken by drawing 700–1500 m^2 of air through a 20×25 cm glass fibre filter (GFF) followed by two polyurethane foam plugs (PUFs). Further experimental details can be inferred from [58]. Enantiomeric ratios of α-HCH in the air samples ranged from 0.93–1.07 ($n=22$) and showed a slight trend with latitude (Fig. 3.35). The ERs were ≥ 1.00 between 40–54°S and ≤ 1.00 from 55–70°S, indicating a depletion of $(-)$-α-HCH at low latitudes and $(+)$-α-HCH at high latitudes. These results clearly show that (1) different enzymatic transformation processes were encountered during the track of the S A Agulhas cruise, and (2) that enzymatic transformation of α-HCH also takes place at higher latitudes in Antarctic regions.

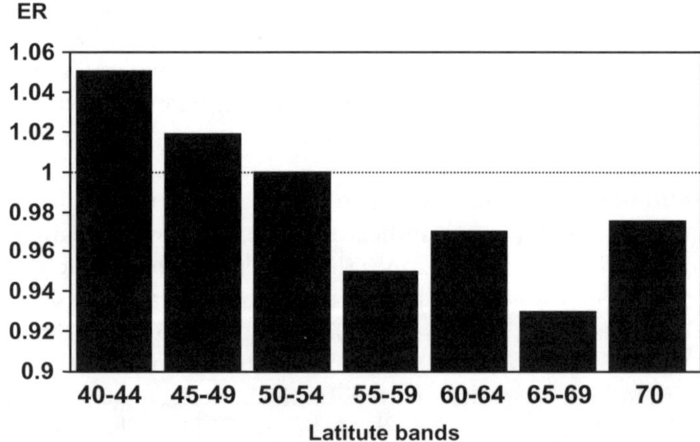

Fig. 3.35. Enantiomeric ratios of α-HCH in air [ER=(+)-/(−)-α-HCH] averaged by 5 degree latitude bands; from [58]

References

1. Ludwig P, Hühnerfuss H, König WA, Gunkel W (1992) Gas chromatographic separation of the enantiomers of marine pollutants. 3. Enantioselective degradation of α-hexachlorocyclohexane by marine microorganisms. Mar Chem 38:13–23
2. Hühnerfuss H, Faller J, König WA, Ludwig P (1992) Gas chromatographic separation of the enantiomers of marine pollutants. 4. Fate of hexachlorocyclohexane isomers in the Baltic and North Sea. Environ Sci Technol 26:2127–2133
3. Cristol JS (1949) The structure of α-benzene hexachloride. J Am Chem Soc 71:1894
4. Vetter W, Klobes U, Luckas B, Hottinger G (1998) Determination of (+/–) elution orders of chiral organochlorines by liquid chromatography with a chiral detector and by enantioselective gas chromatography. J AOAC Int 81:1245–1251
5. Ludwig P (1991) Untersuchungen zum enantioselektiven Abbau von polaren und unpolaren chlorierten Kohlenwasserstoffen durch marine Mikroorganismen. PhD thesis, University of Hamburg, Germany
6. Ludwig P, Gunkel W, Hühnerfuss H (1992) Chromatographic separation of the enantiomers of marine pollutants. 5. Enantioselective degradation of phenoxycarboxylic acid herbicides by marine microorganisms. Chemosphere 24:1423–1429
7. Buser H-R, Müller MD (1995) Isomer and enantioselective degradation of hexachlorocyclohexane isomers in sewage sludge under anaerobic conditions. Environ Sci Technol 29:664–672
8. Faller J, Hühnerfuss H, König WA, Krebber R, Ludwig P (1991) Do marine bacteria degrade α-hexachlorocyclohexane stereoselectively? Environ Sci Technol 25:676–678
9. Faller J, Hühnerfuss H, König WA, Ludwig P (1991) Gas chromatographic separation of the enantiomers of marine organic pollutants. Distribution of α-HCH enantiomers in the North Sea. Mar Pollut Bull 22:82–86
10. Gaul H, Ziebarth U (1983) Methods for the analysis of lipophilic components in water and results about the distribution of different organochlorine components in the North Sea. Dt Hydrogr Z 36:191–212
11. Falconer RL, Bidleman TF, Gregor DJ, Semkin R, Teixeira C (1995) Enantioselective breakdown of α-hexachlorcyclohexane in a small arctic lake and its watershed. Environ Sci Technol 29:1297–1302
12. Falconer RL, Bidleman TF, Gregor DJ (1995) Air-water gas exchange and evidence for metabolism of hexachlorocyclohexanes in Resolute Bay, N.W.T. Sci Total Environ 160/161:65–74
13. Ridal JJ, Bidleman TF, Kerman BR, Fox ME, Strachan WMJ (1997) Enantiomers of α-hexachlorocyclohexanes as tracers of air-water gas exchange in Lake Ontario. Environ Sci Technol 31:1940–1945
14. Bidleman TF, Jantunen LM, Harner T, Wiberg K, Wideman JL, Brice K, Su K, Falconer RL, Aigner EJ, Leone AD, Ridal JJ, Kerman B, Finizio A, Alegria H, Parkhurst WJ, Szeto SY (1998) Chiral pesticides as tracers of air-surface exchange. Environ Pollut 101:1–7
15. Jantunen LM, Bidleman T (1996) Air-water gas exchange of hexachlorocyclohexanes (HCHs) and the enantiomers of α-HCH in arctic regions. J Geophys Res 101:28,837–28,846; Correction: (1997) J Geophys Res 102:19,279–19,282
16. Jantunen LMM, Bidleman TF (1998) Organochlorine pesticides and enantiomers of chiral pesticides in Arctic ocean water. Arch Environ Contam Toxicol 35:218–228
17. Kallenborn R, Hühnerfuss H, König WA (1991) Enantioselective metabolism of (±)-α-hexachlorocyclohexane in organs of the Eider duck. Angew Chem 103:328–329; Angew Chem Int Ed Engl 30:320–321
18. Hühnerfuss H, Kallenborn R (1992) Review. Chromatographic separation of marine organic pollutants. J Chromatogr 580:191–214
19. Möller K, Bretzke C, Hühnerfuss H, Kallenborn R, Kinkel JN, Kopf J, Rimkus G (1994) The absolute configuration of (+)-α-1,2,3,4,5,6-hexachlorocyclohexane and its permeation through the seal blood-brain barrier. Angew Chem 106:911–912; Angew Chem Int Ed Engl 33:882–884

20. Hardt IH, Wolf C, Gehrcke B, Hochmuth DH, Pfaffenberger B, Hühnerfuss H, König WA (1994) Gas chromatographic enantiomer separation of agrochemicals and polychlorinated biphenyls (PCBs) using modified cyclodextrins. J High Resolut Chromatogr 17:859–864

21. Mössner S, Ballschmiter K (1994) Separations of α-hexachlorocyclohexane (α-HCH) and pentachlorocyclohexene (PCCH) enantiomers on a cyclodextrin-phase (Cyclodex-B) by HRGC/ECD. Fresenius J Anal Chem 348:583–589

22. Koske G, Leupold G, Parlar H (1997) Gas chromatographic enantiomer separation of some selected polycyclic xenobiotics using modified cyclodextrins. Fresenius Envir Bull 6:489–493

23. König WA, Icheln D, Runge T, Pfaffenberger B, Ludwig P, Hühnerfuss H (1991) Gas chromatographic enantiomer separation of agrochemicals using modified cyclodextrins. J High Resolut Chromatogr 14:530–536

24. Buser H-R, Müller MD (1992) Enantiomer separation of chlordane components and metabolites using chiral high-resolution gas chromatography and detection by mass spectrometric techniques. Anal Chem 64:3168–3175

25. König WA, Hardt IH, Gehrcke B, Hochmuth DH, Hühnerfuss H, Pfaffenberger B, Rimkus G (1994) Optisch aktive Referenzsubstanzen für die Umweltanalytik. Angew Chem 106:2175–2177; Angew Chem Int Ed Engl 33:2085–2087

26. Oehme M, Kallenborn R, Wiberg K, Rappe C (1994) Simultaneous enantioselective separation of chlordanes, a nonachlor compound, and o,p'-DDT in environmental samples using tandem capillary columns. J High Resolut Chromatogr 17:583–588

27. Vetter W, Klobes U, Luckas B, Hottinger G (1997) Enantiomeric resolution of persistent compounds of technical toxaphene (CTTs) on t-butyldimethylsilylated β-cyclodextrin phases. Chromatographia 45:255–262

28. Karlsson H, Oehme M, Müller L (1996) Enantioselective separation of toxaphene congeners. Results, problems, and challenges. Organohalogen Compd 28:405–409

29. König WA, Gehrcke B, Runge T, Wolf C (1993) Gas chromatographic separation of atropisomeric alkylated and polychlorinated biphenyls using modified cyclodextrins. J High Resolut Chromatogr 16:376–378

30. Vetter W, Klobes U, Luckas B, Hottinger G (1997) Enantiomer separation of selected polychlorinated biphenyls including PCB 144 on $tert$-butyldimethylsilylated β-cyclodextrin. J Chromatogr A 769:247–252.

31. Schurig V, Glausch A (1993) Enantiomer separation of atropisomeric polychlorinated biphenyls (PCBs) by gas chromatography on Chirasil-Dex. Naturwissenschaften 80:468–469

32. Vetter W, Klobes U, Hummert K, Luckas B (1997) Gas chromatographic separation of chiral organochlorines on modified cyclodextrin phases and results in marine biota. J High Resolut Chromatogr 20:85–93

33. Müller MD, Buser H-R, Rappe C (1997) Enantioselective determination of various chlordane components and metabolites using high-resolution gas chromatography with a β-cyclodextrin derivative as chiral selector and electron-capture negative ion mass spectrometry detection. Chemosphere 34:2407–2417

34. Haglund P, Wiberg K (1996) Determination of the gas chromatographic elution sequences of the (+)- and (–)-enantiomers of stable atropisomeric PCBs on Chirasil-Dex. J High Resolut Chromatogr 19:373–376

35. Ellerichmann T, Bergman Å, Franke S, Hühnerfuss H, Jakobsson E, König WA, Larsson C (1998) Gas chromatographic enantiomer separation of chiral PCB methyl sulfones and identification of selectively retained enantiomers in human liver. Fresenius Envir Bull 7:244–257

36. Bergman Å, Ellerichmann T, Franke S, Hühnerfuss H, Jakobsson E, König WA, Larsson C (1998) Gas chromatographic enantiomer separations of chiral PCB methyl sulfones and identification of selectively retained enantiomers in human liver. Organohalogen Compd 35:339–342

37. Gatermann R, Biselli S, Hühnerfuss H, Rimkus GG, Franke S, Hecker M, Kallenborn R, Karbe L, König WA (2001) Synthetic musks in the environment. 2. Enantioselective transformation of HHCB (Galaxolide®), AHTN (Tonalide®), AHDI (Phantolide®), and ATII (Traseolide®) in freshwater fish. Submitted for publication

38. Pfaffenberger B, Hühnerfuss H, Kallenborn R, Köhler-Günther A, König WA, Krüner G (1992) Chromatographic separation of the enantiomers of marine pollutants. 6. Comparison of the enantioselective degradation of α-hexachlorocyclohexane in marine biota and water. Chemosphere 25:719–725

39. Hühnerfuss H, Faller J, Kallenborn R, König WA, Ludwig P, Pfaffenberger B, Oehme M, Rimkus G (1993) Enantioselective and nonenantioselective degradation of organic pollutants in the marine ecosystem. Chirality 5:393–399

40. Hühnerfuss H, Kallenborn R, König WA, Rimkus G (1992) Preferential enrichment of the (+)-α-hexachlorocyclohexane enantiomers in cerebral matter of harbour seals. Organohalogen Compd 10:97–100

41. Mössner S, Spraker TR, Becker PR, Ballschmiter K (1992) Ratios of enantiomers of α-HCH and determination of α-, β-, and γ-HCH isomers in brain and other tissues of neonatal northern fur seals (*Callorhinus ursinus*). Chemosphere 24:1171–1180

42. Müller MD, Schlabach M, Oehme M (1992) Fast and precise determination of α-hexachlorocyclohexane enantiomers in environmental samples using chiral high-resolution gas chromatography. Environ Sci Technol 26:566–569

43. Pfaffenberger B, Hardt I, Hühnerfuss H, König WA, Rimkus G, Glausch A, Schurig V, Hahn J (1994) Enantioselective degradation of α-hexachlorocyclohexane and cyclodiene insecticides in roe-deer liver samples from different regions of Germany. Chemosphere 29:1543–1554

44. Möller K, Hühnerfuss H, Rimkus G (1993) On the diversity of enzymatic degradation pathways of α-hexachlorocyclohexane as determined by chiral gas chromatography. J High Resolut Chromatogr 16:672–673

45. Buser H-R, Müller MD, Rappe C (1992) Enantioselective determination of chlordane components using chiral high-resolution gas chromatography-mass spectrometry with application to environmental samples. Environ Sci Technol 26: 1533–1540

46. Wiberg K, Oehme M, Haglund P, Karlsson H, Olsson M, Rappe C (1998) Enantioselective analysis of organochlorine pesticides in herring and seal from the Swedish marine environment. Mar Pollut Bull, 36:345–353.

47. Buser H-R, Müller MD (1993) Enantioselective determination of chlordane components, metabolites, and photoconversion products in environmental samples using chiral high-resolution gas chromatography and mass spectrometry. Environ Sci Technol 27:1211–1220

48. Zhu J, Norstrom RJ, Muir DCG, Ferron LA, Weber J-P, Dewailly E (1995) Persistent chlorinated cyclodiene compounds in ringed seal blubber, polar bear fat, and human plasma from northern Québec, Canada: identification and concentrations of photoheptachlor. Environ Sci Technol 29:267–271

49. Bethan B, Bester K, Hühnerfuss H, Rimkus G (1997) Bromocyclen contamination of surface water, waste water and fish from northern Germany, and gas chromatographic chiral separation. Chemosphere 34:2271–2280

50. Pfaffenberger B, Hühnerfuss H, Gehrcke B, Hardt I, König WA, Rimkus G (1994) Gas chromatographic separation of the enantiomers of bromocyclen in fish samples. Chemosphere 29:1385–1391

51. Buser H-R, Müller MD (1994) Isomer- and enantiomer-selective analyses of toxaphene components using chiral high-resolution gas chromatography and detection by mass spectrometry/mass spectrometry. Environ Sci Technol 28:119–128

52. Hühnerfuss H, Pfaffenberger B, Gehrcke B, Karbe L, König WA, Landgraff O (1995) Stereochemical effects of PCBs in the marine environment: seasonal variation of coplanar and atropisomeric PCBs in blue mussels (*Mytilus edulis* L.) of the German Bight. Mar Pollut Bull 30:332–340

53. Glausch A, Nicholson GJ, Fluck M, Schurig V (1994) Separation of the enantiomers of stable atropisomeric polychlorinated biphenyls (PCBs) by multidimensional gas chromatography on Chirasil-Dex. J High Resolut Chromatogr 17:347–349

54. Glausch A, Hahn J, Schurig V (1995) Enantioselective determination of chiral 2,2',3,3',4,6'-hexachlorobiphenyl (PCB 132) in human milk samples by multidimensional gas chromatography/electron capture detection and by mass spectrometry. Chemosphere 30:2079–2085

55. Buser H-R, Müller MD, Theobald N (1998) Occurrence of the pharmaceutical drug clofibric acid and the herbicide mecoprop in various Swiss lakes and in the North Sea. Environ Sci Technol 32:188–192

56. Wiberg K, Letcher R, Sandau C, Norstrom R, Tysklind M, Bidleman T (1998) Enantioselective analysis of organochlorines in the Arctic marine food chain: chiral biomagnification factors and relationships of enantiomeric ratios, chemical residues and biological data. Organohalogen Compd 35:371–374

57. Koske G, Leupold G, Angerhöfer D, Parlar H (1998) Multidimensional gas chromatographic enantiomer quantification of some polycyclic xenobiotics in cod liver and fish oils. Organohalogen Compd 35:363–366

58. Jantunen LM, Kylin H, Bidleman TF (1998) Air-water gas exchange of hexachlorocyclohexanes and the enantiomeric ratios of α-HCH in the South Atlantic Ocean and Antarctica. Organohalogen Compd 35:347–350

59. Harner T, Kylin H, Bidleman TF, Strachan WMJ (1998) Air-water gas exchange of hexachlorocyclohexanes (HCHs) and the enantiomers of α-HCH in the Eastern Arctic Ocean. Organohalogen Compd 35:355–358

60. Franke S, Meyer C, Specht M, König WA, Francke W (1998) Chloro-bis-propyl ethers in the Elbe river – isomeric distribution and enantioselective degradation. J High Resolut Chromatogr 21:113–120

61. Miyazaki T, Yamagishi T, Matsumoto M (1985) Isolation and structure of some components in technical grade chlordane. Arch Environ Contam Toxicol 14:475–483

62. Dearth MA, Hites RA (1991) Complete analysis of technical chlordane using negative-ionization mass spectrometry. Environ Sci Technol 25:245–254

63. Fleming I (1988) Grenzorbitale und Reaktionen organischer Verbindungen. VCH Verlagsgesellschaft, Weinheim, Germany

64. Hühnerfuss H, Pfaffenberger B, Rimkus G (1996) Enantioselective transformation and accumulation of cyclodiene pesticides at different trophic levels of marine and terrestrial biota. Organohalogen Compd 29:88–93

65. Karlsson H, Oehme M, Burkow IC, Evenseth A (1997) Determination of enantiomer ratios of chlordane congeners in cod tissues by HRGC/NICI-MS. Technical problems and influence of biological parameters. Organohalogen Compd 31:250–253

66. Bethan B, Bester K, Hühnerfuss H, Rimkus G (1996) Contamination and gas chromatographic chiral separation of bromocyclen in water and fish samples from the River Stör in the northern part of Germany and in some sewage plants' waste water. Organohalogen Compd 28:437–440

67. Arnold SF, Klotz DM, Collins BM, Vonier PM, Guillette LJ Jr, McLachlan JA (1996) Synergistic activation of estrogen receptor with combinations of environmental chemicals. Science 272:1489–1492

68. Garrison AW, Nzengung VA, Avants JK, Ellington J, Wolfe NL (1997) Determining the environmental enantioselectivity of o,p'-DDT and o,p'-DDD. Organohalogen Compd 31:256–261

69. Bidleman TF, Jantunen LM, Falconer RL, Aigner E, Finizio A, Szeto SY, Ridal JJ (1996) ACS Div of Environmental Chem, Preprints of Extended Abstracts 36:243–245.

70. Kallenborn R, Oehme M, Vetter W, Parlar H (1994) Enantiomer selective separation of toxaphene congeners isolated from seal blubber and obtained by synthesis. Chemosphere 28:89–98

71. Buser H-R, Müller MD (1994) Isomeric and enantiomeric composition of different commercial toxaphenes and of chlorination products (+)- and (–)-camphenes. J Agric Food Chem 42:393–400

72. Alder L, Palavinskas R, Andrews P (1996) Enantioselective determination of toxaphene components in fish, monkey adipose tissue from a feeding study and human milk. Organohalogen Compd 28:410–415

73. Vetter W, Klobes U, Luckas B, Hottinger G (1997) Enantioselective determination of toxaphene and other organochlorines on *tert*-butyldimethylsilylated β-cyclodextrin. Organohalogen Compd 33:63–67

74. Vetter W, Schurig V (1997) Enantioselective determination of chiral organochlorine compounds in biota by gas chromatography on modified cyclodextrins. J Chromatogr A774:143–175

75. Andrews P, Vetter W (1995) A systematic nomenclature system for toxaphene congeners. 1. Chlorinated bornanes. Chemosphere 31:3879–3886

76. Oehme M, Kallenborn R (1995) A simple numerical code for polychlorinated compound classes allowing an unequivocal derivation of the steric structure. I. Polychlorinated biphenyls and bornanes. Chemosphere 30:1739–50; reply to comments (1995) 31:3384–3385

77. Nikiforov VA, Tribulovich VG, Karavan, VS (1995) On the nomenclature of toxaphene congeners. Organohalogen Compd 26:393–396

78. Jensen S (1966) Report of a new chemical hazard. New Sci 32:612

79. Waid JS (1986) PCB and the environment. Vol I, CRC Press, Florida, pp 79–100

80. de Boer J (1988) Trends in chlorobiphenyl contents in livers of Atlantic cod (*Gadus morhua*) from the North Sea, 1979–1987. Chemosphere 17:1811–1819

81. Duinker JC, Hillebrand MTJ, Zeinstra T, Boon JP (1989) Individual chlorinated biphenyls and pesticides in tissues of some cetean species from the North Sea and the Atlantic Ocean. Tissue distribution and biotransformation. Aquatic Mammals 15:95–124

82. Kallenborn R, Hühnerfuss H (1993) Vergleich der Verteilung von ausgewählten chlorierten organischen Schadstoffen in Eiderenten (*Somateria mollissima* L.) aus den Einzugsbereichen der Naturschutzgebiete Neuwerk und Oehe-Schleimünde. Seevögel 14:23–31

83. Hühnerfuss H, Pfaffenberger B, Gehrcke B, Karbe L, König WA, Landgraff O (1995) Stereochemical effects of PCBs in the marine environment: seasonal variation of coplanar and atropisomeric PCBs in blue mussels (*Mytilus edulis* L.) of the German Bight. Mar Pollut Bull 30:332–340

84. Kaiser KLE (1974) On the optical activity of polychlorinated biphenyls. Environ Pollut 7:93–101

85. Hühnerfuss H, Pfaffenberger B, Gehrcke B, Karbe L, König WA, Landgraff O (1994) Gas chromatographic enantiomer separation and potential toxicity of atropisomeric polychlorinated biphenyls in marine biota: has the toxic equivalence factor "TEF" concept to be expanded? Organohalogen Compd 21:15–20

86. Haglund P (1995) Isolation of PCB atropisomers for toxicological testing using chiral high-performance liquid chromatography. Organohalogen Compd 23:35–40

87. Schurig V, Glausch A, Fluck M (1995) On the enantiomerization barrier of atropisomeric 2,2',3,3',4,6'-hexachlorobiphenyl (PCB 132). Tetrahedron Asymmetry 6:2161–2164

88. Haglund P (1996) Enantioselective separation of polychlorinated biphenyl atropisomers using chiral high-performance liquid chromatography. J Chromatogr A 724:219–228

89. Haglund P (1996) Isolation and characterization of polychlorinated biphenyl (PCB) atropisomers. Chemosphere 32:2133–2140

90. Blanch GP, Glausch A, Schurig V, Serrano R, González MJ (1996) Quantification and determination of enantiomeric ratios of chiral PCB 95, PCB 132, and PCB 149 in shark liver samples (*C. coelolepis*) from the Atlantic Ocean. J High Resolut Chromatogr 19:392–396

91. Ramos L, Jiménez B, Fernández M, Hernández L, González MJ (1996) Coplanar and chiral PCB determinations using pyrenyl-silica HPLC and HRGC/ECD. Applications to real samples. Organohalogen Compd 27:376–379

92. Müller U, Vetter W, Hummert K, Luckas B (1996) Levels and enantiomeric ratios of oxychlordane and α-HCH in harbour seals (*Phoca vitulina*) and grey seal (*Halichoerus grypus*) from Iceland. Organohalogen Compd 29:118–121

93. Fowler J, Chen L (1990) Practical statistics for field biology. Open University Press, Milton Keynes, UK, pp 130–136

94. Rimkus G, Wolf M (1987) Schadstoffbelastung von Wild aus Schleswig-Holstein. 2. Mitteilung: Rückstände von Dieldrin, Heptachlorepoxid und anderen Cyclodien-Insektiziden im Leberfett von Hasen (*Lepus europaeus* L.) Z Lebensm Unters Forsch 184:308–312

95. Möller K (1993) Untersuchungen zur enantioselektiven Anreicherung von chiralen Schadstoffen im marinen und terrestrischen Ökosystem. Master thesis (Diplomarbeit), University of Hamburg, Hamburg, Germany, p 49

96. Pfaffenberger B (1995) Untersuchungen zur enantioselektiven Anreicherung von chiralen organischen Schadstoffen im marinen und terrestrischen Ökosystem. Dissertation, University of Hamburg, Verlag Shaker, Aachen, Germany, 182 pp

97. Hummert K, Vetter W, Luckas B (1995) Levels of α-HCH, lindane, and enantiomeric ratios of α-HCH in marine mammals from the northern hemisphere. Chemosphere 31:3489–3500

98. Klobes U, Vetter W, Luckas B, Hottinger G (1998) Enantioselective determination of 2-*endo*,3-*exo*,5-*endo*,6-*exo*,8,8,9,10-octachlorobornane (B8–1412) in environmental samples. Organohalogen Compd 35:359–362

99. Vetter W, Luckas B (1998) Analytical artifacts during enantioselective determination of chiral organochlorines with GC/ECNI-MS. Organohalogen Compd 35:367–370

100. Reich S, Jiménez B, Marsili L, Hernández LM, Schurig V, González MJ (1998) Enantiomeric ratios of chiral PCBs in striped dolphins (*Stenella coeruleoalba*) from the Mediterranean Sea. Organohalogen Compd 35:335–338

101. Jensen S, Jansson B (1976) Methyl sulphone metabolites of PCB and DDE. Ambio 5:257–260

102. Kato Y, Haraguchi K, Kawashima M, Yamada S, Isogai M, Masuda Y, Kimura R (1995) Characterization of hepatic microsomal cytochrome P-450 from rats treated with methyl sulfonyl metabolites of polychlorinated biphenyl congeners. Chem Biol Interact 95:269–278

103. Kato Y, Haraguchi K, Tomiyasu K, Saito H, Isogai M, Masuda Y, Kimura R (1997) Structure-dependent induction of CYP2B1/2 by 3-methylsulfonyl metabolites of polychlorinated biphenyl congeners in rats. Environ Toxicol Pharmacol 3:137–144

104. Larsson C, Ellerichmann T, Franke S, Athanasiadou M, Hühnerfuss H, Bergman Å (1999) Enantiomeric separation of chiral methylsulphonyl PCB congeners in liver and adipose tissue from rats dosed with A50. Organohalogen Compd 40: 427–430

105. Hühnerfuss H, Kallenborn R, König WA, Rimkus G (1992) Preferential enrichment of the (+)-α-hexachlorocyclohexane in cerebral matter of harbour seals. Organohalogen Compd 10:97–100

106. Hühnerfuss H, Kallenborn R, Möller K, Pfaffenberger B, Rimkus G (1993) Enzymatic enantioselective degradation of α-hexachlorocyclohexane in different tissues of marine and terrestrial biota as determined by chiral gas chromatography. Proc 15th Int Symp on Gas Chromatogr, Riva, pp 576–581

107. Möller K (1993) Untersuchungen zur enantioselektiven Anreicherung von chiralen Schadstoffen im marinen und terrestrischen Ökosystem. Master thesis (Diplomarbeit), University of Hamburg, Hamburg, Germany, p 49

108. Portig J, Stein K, Vohland HW (1989) Preferential distribution of α-hexachlorocyclohexane into cerebral white matter. Xenobiotica 19:123–130

109. Gatermann R, Biselli S, Hühnerfuss H, Rimkus GG, Hecker M, Karbe L (2001) Synthetic musks in the environment. 1. Species-dependent bioaccumulation for HHCB, AHTN, musk xylene and musk ketone in freshwater fish. Submitted

110. Itrich NR, Simonich SL, Federle TW (1998) Biotransformation of the polycyclic musk, HHCB, during sewage treatment. Poster Presentation at the SETAC 19th Annual Meeting, 14–15 November, Charlotte, NC, USA
111. Hühnerfuss H, Gatermann R, Biselli S, Rimkus GG, Hecker M, Kallenborn R, Karbe L (1999) Enantioselective transformation of polycyclic musks in aquatic biota. Organohalogen Compd 40:401–404
112. Gatermann R, Hühnerfuss H, Biselli S, Rimkus GG, Hecker M, Kallenborn R, Karbe L (1999) Species-dependent enantioselective transformation of HHCB, AHTN and ATII in crucian carp (*Carassius carassius*) and tench (*Tinca tinca*). Organohalogen Compd 40:595–598
113. Biselli S, Dittmer H, Gatermann R, Kallenborn R, König WA, Hühnerfuss H (1999) Separation of HHCB, ATII, AHDI and DPMI by enantioselective capillary gas chromatography and preparative separation of HHCB and ATII by enantioselective HPLC. Organohalogen Compd 40:599–602
114. Ohloff G (1990) Riechstoffe und Geruchssinn – Die molekulare Welt der Düfte, Springer-Verlag, Berlin, Heidelberg, New York, pp 195–209
115. Rimkus GG (1999) Polycyclic musk fragrances in the aquatic environment. Toxicol Lett 111:37–56
116. Yamagishi T, Miyazaki T, Horii S, Kaneko S (1981) Identification of musk xylene and musk ketone in freshwater fish collected from the River Tama. Tokyo Bull Environ Contam Toxic 26:656–662
117. Yamagishi T, Miyazaki T, Horii S, Akiyama K (1983) Synthetic musk residues in biota and water from the River Tama and Tokyo Bay. Arch Environ Contam Toxicol 12:83–89
118. Gatermann R, Hühnerfuss H, Rimkus G, Wolf MS, Franke S (1995) The distribution of nitrobenzene and other nitroaromatic compounds in the North Sea. Mar Pollut Bull 30:221–227
119. Eschke HD, Traud J, Dibowski H-J (1994) Analytik und Befunde künstlicher Nitromoschussubstanzen in Oberflächen- und Abwässern sowie Fischen aus dem Einzugsgebiet der Ruhr. Vom Wasser 83:373–383
120. Rimkus GG, Wolf M (1995) Nitro musk fragrances in biota from fresh-water and marine environment. Chemosphere 30:641–651
121. Rimkus GG, Rimkus B, Wolf M (1994) Nitro musks in human adipose tissue and breast milk. Chemosphere 28:421–432
122. Liebl B, Ehrenstorfer S (1993) Nitro musks in human milk. Chemosphere 27:2253–2260
123. Müller S, Schmid P, Schlatter C (1996) Occurrence of nitro and non-nitro benzenoid musk compounds in human adipose tissue. Chemosphere 33:17–28
124. Verordnung (EG) Nr. 143/97 der Kommission vom 27. Januar 1997 zur Festlegung der dritten Prioritätsliste gemäß der Verordnung (EWG) Nr. 793/93, Amtsblatt der Europäischen Gemeinschaften Nr. l 25/13
125. OSPAR convention for the protection of the marine environment of the North-East Atlantic ad hoc working group on the development of a dynamic selection and prioritisation mechanism for hazardous substances, Berlin 14–16.9.1998
126. Rimkus GG, Butte W, Geyer HJ (1997) Critical considerations on the analysis and bioaccumulation of musk xylene and other synthetic nitro musks in fish. Chemosphere 35:1497–1507
127. Gatermann R, Hühnerfuss H, Rimkus G, Attar A, Kettrup A (1998) Occurrence of musk xylene and musk ketone metabolites in the aquatic environment. Chemosphere 36:2535–2547
128. Rimkus GG (1998) Synthetic musk fragrances in human fat and their potential uptake by dermal resorption. In: Frosch PJ, Johansen JD, White IR (eds) Fragrances. Beneficial and adverse effects. Springer-Verlag, Berlin, Heidelberg, New York, p 136
129. van de Plassche EJ, Balk F (1997) Environmental risk assessment of the polycyclic musks AHTN and HHCB according to the EU-TGD. RIVM-Report No. 6011503008, Bilthoven

130. Eschke H-D, Traud J, Dibowski H-J (1994) Untersuchungen zum Vorkommen polycyclischer Moschus-Duftstoffe in verschiedenen Umweltkompartimenten – Nachweis und Analytik mit GC/MS in Oberflächen-, Abwässern und Fischen. UWSF-Z Umweltchem Ökotox 6:183–189

131. Eschke H-D, Dibowski H-J, Traud J (1995) Untersuchungen zum Vorkommen polycyclischer Moschus-Duftstoffe in verschiedenen Umweltkompartimenten. UWSF-Z Umweltchem Ökotox 7:131–135

132. Rimkus G, Brunn H (1996) Synthetische Moschusduftstoffe – Anwendung, Anreicherung in der Umwelt und Toxikologie. 1. Herstellung, Anwendung, Vorkommen in Lebensmitteln, Aufnahme durch den Menschen. Ernährung-Umschau 43:442–449

133. Brunn H, Rimkus G (1997) Synthetische Moschusduftstoffe – Anwendung, Anreicherung in der Umwelt und Toxikologie. 2. Toxikologie der synthetischen Moschusduftstoffe und Schlußfolgerungen. Ernährung-Umschau 44:4–9

134. Franke S, Meyer C, Heinzel N, Gatermann R, Hühnerfuss H, Rimkus G, König WA, Francke W (1999) Enantiomeric composition of the polycyclic musks HHCB and AHTN in different aquatic species. Chirality 11:795–801

135. Frater G, Müller U, Bajgrowicz JA, Petrzilka M (1995) Musky and odorless diastereomers of galaxolide. In: Baser KHC (ed) Flavours, fragrances and essential oils, Proc 13th Int Congress of flavours, fragrances and essential oils, Istanbul, Turkey, October 15–19

136. Müller S, Schmid P, Schlatter C (1996) Occurrence of nitro and non-nitro benzenoid musk compounds in human adipose tissue. Chemosphere 33:17–28

137. König WA (2000) Chirality in the natural world – odours and tastes. In: Wainer IW, Laugh J (eds) 150 Years after Pasteur. Kloever-Verlag, in press

138. Baydar A, Petrzilka M, Schott MP (1993) Olfactory threshold for androstene and Galaxolide: sensitivity, insensitivity and specific anosmia. Chem Senses 18:661–668

139. Gower DB, Ruparelia BA (1993) Olfaction in humans with special reference to odorous 16-androstenes: their occurrence, perception and possible social, psychological and sexual impact. J Endocrinol 137:167–187

140. Gilbert AN, Wysocki CJ (1991) Quantitative assessment of olfactory experience during pregnancy. Psychosomatic Medicine 53:693–700

141. Buck L, Axel R (1991) A novel multigene family may encode odorant receptors: a molecular basis for odor recognition. Cell 65:175–187

142. Stacey NE, Sorensen PW, Cardwell JR (1993) Hormonal pheromones: recent developments and potential applications in aquaculture. Coastal and Estuarine Studies, Berlin, pp 227–239

143. Bjerselius R, Olsén KH (1993) The olfactory sensitivity of crucian carp (*Carassius carassius*) and goldfish (*Carassius auratus*) to 17,20b-dihydroxy-4-pregnen-3-one and prostaglandin F2. Chem Senses 18:427–436

144. Müller MD, Buser H-R (1995) Environmental behaviour of acetamide pesticide stereoisomers. 2. Stereo- and enantioselective degradation in sewage sludge and soil. Environ Sci Technol 29:2031–2037

145. Vetter W, Bartha R, Stern G, Tomy G (1998) Enantioselective determination of two major compounds of technical toxaphene in Canadian lake sediment cores from the last 60 years. Organohalogen Compd 35:343–346

146. Vetter W, Bartha R, Stern G, Tomy G (1999) Enantioselective determination of two persistent chlorobornane congeners in sediment from a toxaphene treated Yukon lake. Environ Toxicol Chem 18:2775–2781

147. Fingerling G, Hertkorn N, Parlar H (1996) Formation and spectroscopic investigation of two hexachlorobornanes from six environmentally relevant toxaphene components by reductive dechlorination in soil under anaerobic conditions. Environ Sci Technol 30:2984–2992

148. Rappe C, Haglund P, Buser H-R, Müller MD (1997) Enantioselective determination of chiral chlorobornanes in sediments from the Baltic Sea. Organohalogen Compd 31:233–237

149. Benická E, Novakovski R, Hrouzek J, Krupcik J (1996) Multidimensional gas chromatographic separation of selected PCB atropisomers in technical formulations and sediments. J High Resolut Chromatogr 19:95–98

150. Ariëns EJ, van Rensen JJS, Welling W (1988) Stereoselectivity of pesticides. Biological and chemical problems. Elsevier, Amsterdam

151. Ramos Tombo GM, Belluš D (1991) Chiralität und Pflanzenschutz. Angew Chem 103:1219–1241

152. Buser H-P, Francotte E (1996) Stereoselective analysis in crop protection. In: Ahuja S (ed) Chiral separations: applications and technology. ACS Professional Reference Books, Washington, chap 5, pp 93–138

153. Bidleman TF, Jantunen LM, Harner T, Wiberg K, Wideman J, Falconer RL, Aigner EJ, Lenone AD, Ridal JJ, Kerman B, Parkhurst W, Finizio A, Szeto SY (1997) Chiral pesticides as tracers of air-surface exchange. Organohalogen Compd 31:238–243

154. Wiberg K, Jantunen LM, Harner T, Wideman JL, Bidleman TF, Brice K, Su K, Falconer RL, Leone AD, Parkhurst W, Alegria H (1997) Chlordane enantiomers as source markers in ambient air. Organohalogen Compd 33:209–213

155. Bidleman TF, Jantunen LM, Wiberg K, Harner T, Brice KA, Su K, Falconer RL, Leone AD, Aigner EJ, Parkhurst WJ (1998) Soil as a source of atmospheric heptachlor epoxide. Environ Sci Technol 32:1546–1548

156. Falconer R, Leone A, Bodnar C, Wiberg K, Bidleman T, Jantunen L, Harner T, Parkhurst W, Alegria H, Brice K, Su K (1998) Using enantiomeric ratios to determine sources of chlordane to ambient air. Organohalogen Compd 35:331–334

157. Aigner E, Leone A, Falconer R (1998) Concentrations and enantiomers of pesticides in soil from the U.S. cornbelt. Environ Sci Technol 32:1162–1168

158. Bidleman T, Alegria H, Ngabe B, Green C (1998) Trends of chlordane and toxaphene in ambient air of Columbia, South Carolina. Atmos Environ 32:1849–1856

159. Finizio A, Bidleman TF, Szeto SY (1998) Emission of chiral pesticides from an agricultural soil in the Fraser Valley, British Columbia. Chemosphere 36:345–355

160. Falconer RL, Bidleman TF, Szeto SY (1997) Chiral pesticides in soils of the Fraser Valley, British Columbia. J Agric Food Chem 45:1946–1951

161. Hühnerfuss H, Garrett WD (1981) Experimental sea slicks: their practical applications and utilization for basic studies of air-sea interactions. J Geophys Res 86:439–447.

162. Ulrich EM, Hites RA (1998) Enantiomeric ratios of chlordane-related compounds in air near the Great Lakes. Environ Sci Technol 32:1870–1874

163. Kallenborn R, Hartwig E, Hühnerfuss H (1994) Vergleich der Nahrung der Eiderenten (*Somateria mollissima* L.) aus unterschiedlichen Rastgebieten. Seevögel 15:31–37.

164. Lewis DL, Garrison AW, Wommack KE, Whittemore A, Steudler P, Melillo J (1999) Influence of environmental changes on degradation of chiral pollutants in soils. Nature:401:898–901

165. de Geus HJ, Baycan-Keller R, Oehme M, de Boer J, Brinkman UATh (1998) Enantiomer ratios of bornane congeners in biological samples using heart-cut gas chromatography with an enantioselective column. J High Resolut Chromatogr 21:39–46

166. Karlsson H, Oehme M, Scherer G (1998) Structure elucidation of U82, a major octachloro chlordane congener present at higher trophic levels in marine food webs and in human tissue. Organohalogen Compd 35:51–55

167. Karlsson H, Oehme M, Scherer G (1999) Isolation of the chlordane compounds U82, MC5, MC7, and MC8 from technical chlordane by HPLC including structure elucidation of U82 and determination of ECD and NICI-MS response factors. Environ Sci Technol 33:1353–1358

168. Weigel S (1998) Entwicklung einer Methode zur Extraktion organischer Spurenstoffe aus großvolumigen Wasserproben mittels Festphasen. Master thesis (Diplomarbeit), University of Hamburg, p 86

169. Buser H-R, Poiger T, Müller MD (1999) Occurrence and environmental behaviour of the chiral pharmaceutical drug ibuprofen in surface waters and in wastewater. Environ Sci Technol 33:2529–2535

170. Ternes T (1998) Occurrence of drugs in German sewage treatment plants and rivers. Water Res 32:3245–3260
171. Harju MT, Haglund P (1999) Determination of rotational barriers of atropisomeric polychlorinated biphenyls. J Anal Chem 364:219–223
172. Ellerichmann T (2000) Analyse chiraler PCB und ihrer Methylsulfonyl-Metaboliten in Biotaproben mit Hilfe enantioselektiver Gas- und Flüssigkeitschromatographie unter Verwendung von Cyclodextrinderivaten. PhD thesis, University of Hamburg, Verlag Shaker, Aachen, Germany, p 118
173. Biselli S (2001) PhD thesis, University of Hamburg, in preparation
174. Pfeiffer CC (1956) Optical isomerism and pharmacological action, a generalization. Science 124:29–41
175. Buser H-R, Müller MD (1995) Isomer-selective and enantiomer-selective determination of DDT and related compounds using chiral high-resolution gas chromatography/mass spectrometry and chiral HPLC. Anal Chem 67:2691–2698
176. Rimkus GG, Wolf M (1993) Contaminants in fish from aquaculture. 2. Musk xylene and musk ketone contaminants in fish. Dts Lebensmittel-Rundschau 89:171–175
177. Vetter W (1993) Toxaphene – Theoretical aspects of the distribution of chlorinated bornanes including symmetry aspects. Chemosphere 26:1079–1084
178. BidlemanTF, Falconer RL (1999) Enantiomer ratios for apportioning two sources of chiral compounds. Environ Sci Technol 33:2299–2301

4 Enantioselective Toxic and Ecotoxic Effects of Drugs and Environmental Pollutants

4.1
Differential Toxic Effects of Drug Enantiomers and the Role of Enantioselective Chromatography

Over the last two decades interest in the pharmacodynamic and pharmacokinetic properties of the enantiomers of racemic chiral drugs has increased considerably and has become an issue of concern to both the pharmaceutical industry and the regulatory authorities [1]. This current interest has been stimulated not only in part by the "Thalidomide case" (marketed under the trade name Contergan® in Germany), but also by the recent advances in stereoselective synthesis and stereospecific analysis, particularly in the area of chromatography. As a result of the technological advances in the latter areas it is likely that new drug development will concentrate on single stereoisomers and the development of racemic mixtures will require scientific justification. In the current regulatory climate there appears to be a continuing requirement for enantiospecific analysis and thus modern drug development presents a considerable challenge for the pharmaceutical analyst and separation scientist [1]. As a consequence, enantiospecific analytical methodologies are required throughout the drug discovery/development process. However, in the present volume, we cannot cover this rapidly developing field. The reader interested in these aspects should refer to [2–5]. Herein, we confine ourselves to enantioselective toxic and ecotoxic effects as observed for chiral drugs. This aspect appears to be of increasing concern because, as shown by the recent investigations by Weigel, drugs are being found in the aquatic environment in larger concentrations than previously expected [6].

The "Thalidomide case" can only be appreciated by taking into account the former and current test procedures which have to be carried out prior to bringing a new drug to the market. Actually, the sedative thalidomide [(R,S)-N-(2,6-dioxo-3-piperidyl)phthalimide; Fig. 4.1] was a dramatic example of the potential limitations of toxicological test systems. Although numerous animal tests were performed, they failed to predict the obvious teratogenic effects on unborn children, which led to the well-known tragedy in the early 1960s in Western Europe [7]. At that time, the drug was sold and used as a racemate. First tests car-

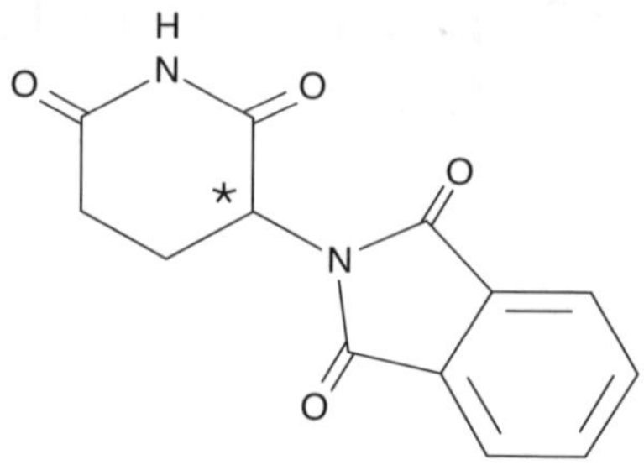

Fig. 4.1. Molecular structure of thalidomide (Contergan®)

ried out immediately after the first reports on its teratogenic effect indicated that only the S-enantiomer is responsible for this effect on unborn children [7]. However, recent re-investigations supplied the surprising result that the presence of both enantiomers may give rise to comparable effects, because the R-enantiomer, which possesses the tranquillising properties, under physiological conditions interconverts into the teratogenic S-enantiomer ([8,55] and literature cited therein). The racemisation appears shortly after application, which makes clear that data concerning the biological properties of a single enantiomer should be viewed with some caution. However, due to the early suspicions of the enantiomer-specific teratogenic effect induced by thalidomide clearly defined regulatory restrictions in terms of enantioselective toxicity and tests were implemented in official rules for chiral drug control which in turn require reliable enantioselective analyses.

It is worth noting that thalidomide is currently being discussed as a "new" drug [9]. The U.S. Food and Drug Administration (FDA) has issued an "approved" letter to its manufacturer in the U.S.A., and this company is now starting to market thalidomide, brand named "Synovir", for only the narrow indication of *Erythema nodosum leprosum*, a leprosy complication with no effective alternative therapy. Clinical data submitted to the FDA has shown that, under tightly controlled medical supervision, benefits outweigh risks of thalidomide therapy for leprosy. It has also been shown that thalidomide is effective against *aphthous ulcers* in patients with HIV infection. Thus, thalidomide may become a "miracle" drug for such diseases as AIDS (AIDS-associated *cachexia*, wasting of the body), but it could also be used to treat *lupus erythematosus, rheumatoid arthritis, multiple sclerosis,* and *tuberculosis.* It may also find applications in *brain* and *prostate cancers, ophthalmic macular degeneration,* and *diabetic retinopathy* [9]. Within the last year, the FDA has approved its use with strict safeguards to protect women who are or could become pregnant.

As early as 1986, enantioselective opioid-hallucinogen interactions were reported for N,N-dimethyltryptamine and lysergic acid N,N-diethylamide (LSD; Fig. 4.2), which induced a model behavioural state in rats [10]. In adult male rats trained on a positive reinforcement fixed ratio of four (FR4: a reward of 0.01 mL sugar-sweetened milk earned on every fourth bar press), N,N-dimethyltryp-

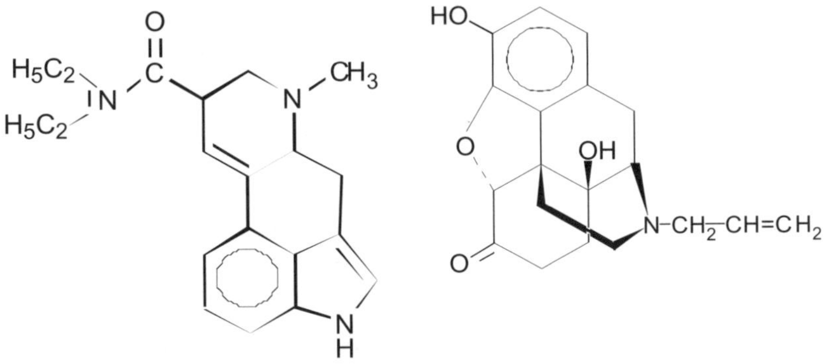

Fig. 4.2. *Left*: molecular structure of LSD (lysergic acid *N,N*-diethylamide); *right*: the opioid antagonist naloxone (17-allyl-4,5α-epoxy-3,14-dihydroxy-6-morphinanone)

tamine, in concentrations of 3.2 and 10.0 mg/kg body weight, disrupted established food-rewarded FR4 bar-pressing behaviour in a dose-related fashion. Predetermined behaviourally ineffective doses of opioid agonists showed selective biphasic effects against *N,N*-dimethyltryptamine and LSD. Low doses antagonised the effects of both hallucinogens, whereas larger doses enhanced their effects. In contrast to the antagonistic effects of low doses of opioid agonists, (–)-naloxone (i.e., 17-allyl-4,5α-epoxy-3,14-dihydroxy-6-morphinanone, Fig. 4.2) enhanced the effects of *N,N*-dimethyltryptamine and LSD. Potentiation of *N,N*-dimethyltryptamine-induced behavioural disruption was attributed to a stereospecific opioid antagonist effect of (–)-naloxone, while the (+)-naloxone enantiomer failed to potentiate the effects of *N,N*-dimethyltryptamine. Thus, behavioural fixed patterns can also help to elucidate stereochemical effects on physiological processes.

4.2
Differential Toxic Effects of Chiral Environmental Pollutants and the Role of Enantioselective Chromatography

In this section, emphasis will be placed on investigations that focus on differential toxic effects of pesticides and other chiral environmental pollutants. With regard to the problem related to "chirality and crop protection", the reader should refer to the comprehensive reviews published by Ariëns et al. [11], Ramos Tombo and Belluš [12], and Buser and Francotte [13].

Several microbiological assays have been developed for the effect testing of chiral pesticides, e.g., an *in vitro* test suitable for transformation studies of four *s*-triazine herbicides (atrazine, terbuthylazine, ametryne and terbutryne) using liver microsomes from rats, pigs and humans [14]. Species-specific stereoselective formation of a new chiral isopropylhydroxylated metabolite from atrazine (Fig. 4.3) was investigated using chiral HPLC techniques and UV-detection with a binary mobile phase containing acetonitrile and water (0.5% phosphoric acid)

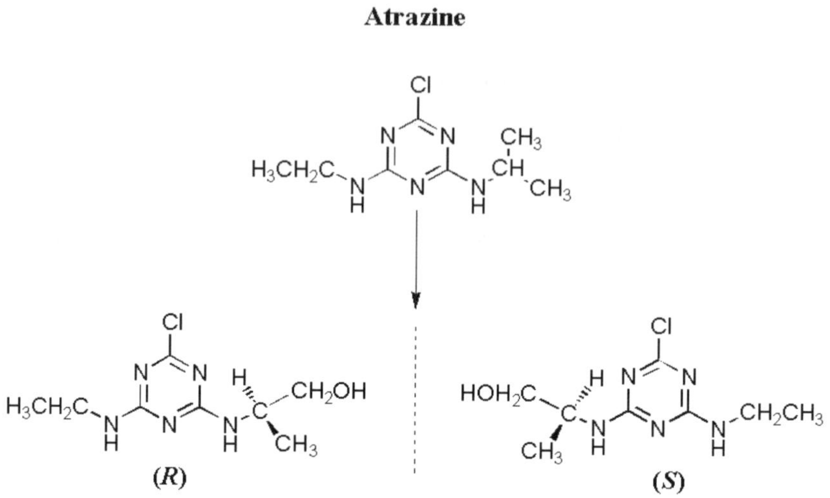

Atrazine

Fig. 4.3. Transformation of atrazine and the molecular structure of the chiral metabolite 2-chloro-4-ethylamino-6-(1-hydroxy-2-methylethyl-2-amino)-1,3,5-triazine; from [14]

for the analysis of atrazine and terbuthylazine as well as acetonitrile and 50 mM potassium phosphate buffer (pH 3.5) for the analysis of ametryne and terbutryne, including metabolites. In addition, the alkylhydroxylated metabolites were derivatised with N-methyl-N-($tert$-butyldimethylsilyl)trifluoroacetamide prior to analysis by cGC/MS. Standards of the metabolites were prepared by Lang et al. [14]. As a result of the transformation of the prochiral atrazine via hydroxylation, a chiral metabolite, the 1-hydroxyisopropylatrazine (abbreviated M1-At), is formed. The distribution of the enantiomers (R/S)-M1-At was found to be different in rat, pig and human liver tissues. The R-enantiomer dominated in human tissues, whereas the S-enantiomer prevailed in rat tissues. The authors reported a species-dependent enantioselective formation of this metabolite, with S/R ratios of 76:24 in rats, 49:51 in pigs, and 28:72 in humans [14]. The authors argued that no species is able to produce only one enantiomer selectively. Therefore, they suggested that differences in the asymmetric induction of the catalytic reaction on the active centre of the enzyme results in different expressions of the stereoselective enzymes which, in turn, lead to the observed characteristic changes in the enantiomers mentioned above. However, this hypothesis has yet to be verified by supporting experiments.

Polychlorocycloalkane and pyrethroid insecticides were shown to inhibit γ-aminobutanoic acid (GABA) dependent chloride flux into mouse brain vesicles [15]. For the cyclodienes, inhibition of chloride uptake was closely correlated with both mammalian toxicity and the ability to displace the binding of ^{35}S $tert$-butylbicyclophosphorothionate. The pyrethroid insecticide induced 50% inhibition of GABA dependent chloride uptake at 25 μM, but the extent of inhibition was not increased at higher concentrations. In addition, the non-toxic enantio-

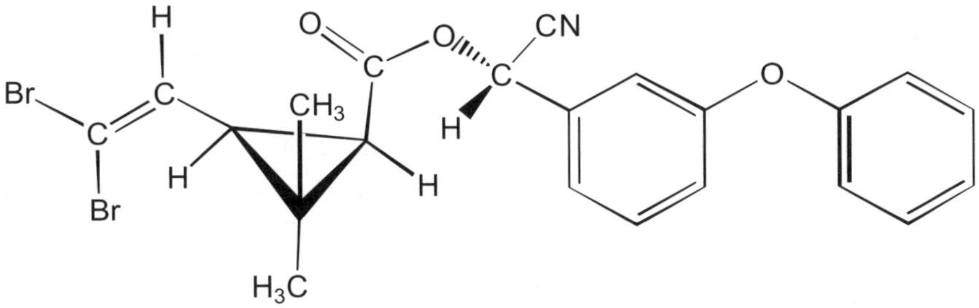

Fig. 4.4. Molecular structure of deltamethrin; from [15]

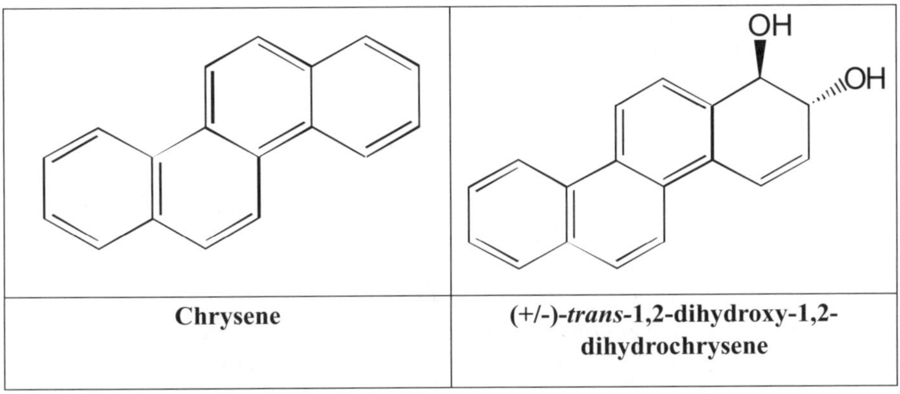

| Chrysene | (+/-)-*trans*-1,2-dihydroxy-1,2-dihydrochrysene |

Fig. 4.5. Chrysene and its metabolite (+/–)-*trans*-1,2-dihydroxy-1,2-dihydrochrysene; from [16]

mer of deltamethrin, the 1S-deltamethrin (Fig. 4.4), produced dose-dependent inhibition in the chloride flux assay with a potency about tenfold less than racemic deltamethrin. These results allow the conclusion that the inhibitory activity of the 1S-deltamethrin on ^{36}Cl uptake demonstrates the stereoselectivity of the GABA receptor ionophore complex for enantiomers. Furthermore, these results demonstrate the utility of this new assay to identify potential hazardous compounds that act at the GABA receptor ionophore complex, which is assumed to represent the principal site of neurotoxic action at least for cyclodiene and pyrethroid insecticides. It is worth noting that the authors included in their assumption that for deltamethrin and 1S-deltamethrin a stereoselective influence on the sodium channel in the sense of a structure/toxicity relationship can be stated.

As early as 1983, an interesting feature was revealed for chrysene, a polyaromatic hydrocarbon (PAH) [16]. The application of (–)-*trans*-1,2,-dihydroxy-1,2-dihydrochrysene (Fig. 4.5) had twice the tumor-initiating activity on mouse skin while the (+)-enantiomer had no significant tumor-initiating activity compared with a standard tumor promoter (tetradecanoylphorbol 13-acetate).

4.3
Acute and Chronic Toxicity of Chiral Environmental Pollutants

Differential effects of the α-HCH enantiomers on cytotoxicity and growth stimulation in primary rat hepatocytes have been observed by Möller et al. [17]. The *cytotoxic effect* was determined as a parameter for the *acute toxicity* of α-HCH, while the *growth stimulation* may be associated with the *chronic toxicity*, e.g., tumor promotion.

For the cytotoxicity assay hepatocytes were cultured for 24 h in the presence of (+)- or (−)-α-HCH. After a change of medium, cells were incubated with neutral red (20 mg/mL) for 3 h. Then dye uptake into the lysosomes of intact cells was determined. The proportion of dead cells was obtained by counting at least 400 cells in five randomly chosen fields of the culture dishes microscopically. In general, the neutral red cytotoxicity assay is used on the following basis: a reduced dye retention in lysosome cells indicates a toxic effect of a respective test compound. However, in the present case, up to concentrations of 3×10^{-4} M of the α-HCH enantiomers no toxic effect could be detected with the neutral red assay. Unexpectedly, an increased adsorption of dye was found although a dose-dependent reduction of all viability was observed microscopically.

Therefore, the numbers of viable and dead cells were counted microscopically. It turned out that a concentration of 3×10^{-4} M (+)-α-HCH leads to a mortality of 100% while, at the same concentration of (−)-α-HCH, 75% dead cells were observed (Fig. 4.6). At lower concentration, i.e., 1×10^{-4} M, no cytotoxic effect of the (−)-enantiomer of α-HCH was found, whereas the (+)-enantiomer still showed

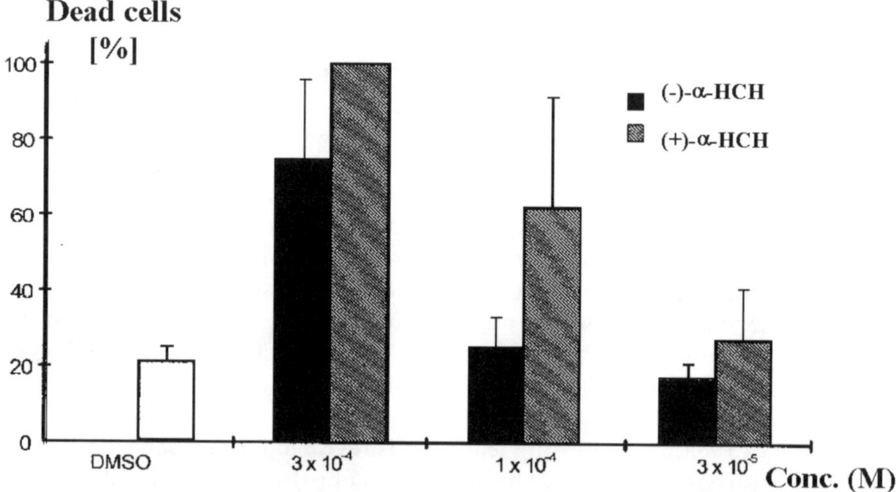

Fig. 4.6. Mortality in primary cultures of rat hepatocytes as a function of (+)-α-HCH and (−)-α-HCH concentrations. Data represent the mean $\pm$ SD (triplicate cultures from 3 male animals; from [17]); DMSO = Control

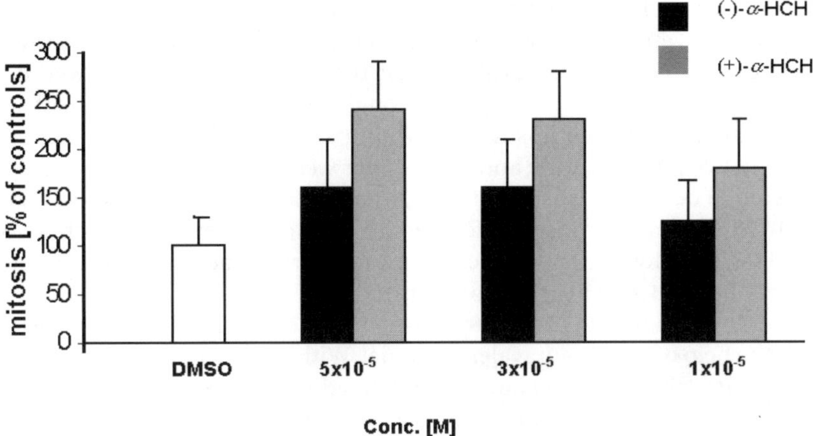

Fig. 4.7. Stimulation of mitosis in primary rat hepatocytes by $(+)$-α-HCH and $(-)$-α-HCH. The DMSO controls were set at 100%. Data represent the mean $\pm$ SD (triplicate cultures of 4 male rats; from [17])

a significant mortality of 62%. These results make evident that caution has to be applied when inferring cytotoxic effects exclusively from the neutral red assay.

The *mitotic rates* of primary rat hepatocytes were stimulated by both enantiomers at concentrations of 5×10^{-5} and 3×10^{-5} M compared with DMSO controls (Fig. 4.7). In the presence of $(+)$-α-HCH (5×10^{-5} M) significantly higher rates of mitosis (factor 2.4) compared with the stimulation of $(-)$-α-HCH (factor 1.7) were observed. At a concentration of 1×10^{-5} M the $(+)$-enantiomer stimulated the mitosis whereas the $(-)$-enantiomer was ineffective.

In summary, a significant difference between the α-HCH enantiomers was observed concerning the cytotoxic effect as well as the growth stimulation. In both assays the $(+)$-α-HCH was more effective than the $(-)$-enantiomer. It may be concluded from these results that an enantioselective enrichment of $(-)$-α-HCH is associated with a lower risk factor than the accumulation of $(+)$-α-HCH. Thus, the quantitative determination of α-HCH as a mixture of both enantiomers may lead to an overestimation of the toxicity in those animal tissues, e.g., the liver of sheep [18] and roe-deer [19], in which the more toxic $(+)$-enantiomer is preferentially degraded.

4.4
Enantioselective Toxic Effects of Other Industrial Chiral Chemicals

In the last few years, increasing attention has been paid to the relationship between the conformation of polychlorinated biphenyls (PCBs) and their impact on marine and terrestrial biota [20–27]. In particular, emphasis has been placed on the toxicity of so-called "coplanar" PCBs, i.e., congeners that are able to attain a coplanar conformation of the two aryl systems due to lack of *ortho*-chlorine substituents. This holds for non-*ortho*- and mono-*ortho*-substituted PCBs. Recent "quantitative-structure-activity relationship (QSAR)" studies have revealed that the non-*ortho* coplanar PCBs exhibit the highest toxicity followed by the moderately toxic mono-*ortho* coplanar congeners, while the di-*ortho*-substituted PCBs turned out to be less toxic [26]. In particular, the coplanar congeners possessing two *para*-chlorine and at least one *meta*-chlorine substituent may be considered to be largely isosters of 2,3,7,8-tetrachlorodibenzo-*p*-dioxin (2,3,7,8-TCDD) [28] and, as a consequence, dioxin-like toxic responses, such as body weight loss, porphyria, teratogenesis, and endocrine and productive malfunctions, may be induced in various organisms [22].

The toxic potency of PCB mixtures in environmental samples is being estimated with the help of a toxic equivalency model. Within the scope of this model, toxic equivalency factors (TEFs) have been defined for the coplanar PCB congeners that induce aryl hydrocarbon hydroxylase (AHH) activity in a similar manner to 2,3,7,8-TCDD. Each of these PCBs has been assigned a TEF value based on its toxicity relative to 2,3,7,8-TCDD, which has by definition a TEF of 1 [26].

A risk assessment of the toxic potential of atropisomeric PCBs is difficult, because the present concept of TEFs is confined to dioxin-like PCBs, as outlined above, and thus not applicable to atropisomeric PCBs. However, some basic insights, which may give rise to an alternative concept that will include atropisomeric PCBs, can be inferred from the pioneering study published by Rodman et al. [29]. The authors investigated the atropisomers of PCB88, PCB139, and PCB197 in chick embryo hepatocyte cultures to determine whether chirality plays a role in the recognition events associated with the induction of cytochrome P-450 and the accumulation of uroporphyrin (URO). Concentration-related induction of cytochrome P-450 content, ethoxyresorufin-*O*-deethylase (EROD) and benzphetamine *N*-demethylase (BPDM) activities were measured.

Rodman et al. [29] determined a rank order of potency for total cytochrome P-450 induction of PCB139>PCB197≥PCB88. The (+)- and (–)-enantiomers of PCB88 and of PCB197 were of equal potencies as inducers of cytochrome P-450, whereas potency of the (+)-enantiomer of PCB139 was greater than of the corresponding (–)-enantiomer.

PCB139 was a much more potent inducer of EROD activity than was either PCB88 or PCB197. Furthermore, EROD activity was induced to a much greater extent by the (+)-enantiomers of all compounds, with the (–)-enantiomers of PCB88 and PCB197 being inactive. As coplanar PCBs are also able to induce EROD activity (see [30] and literature cited therein), it has been suggested by Hühnerfuss et al. [31] that this parameter can be used to infer total toxic potencies, including both coplanar and atropisomeric PCBs. Particularly encouraging are the significantly different responses of the EROD activity to the respective enantiomers of atropisomeric PCBs.

According to Rodman et al. [29], BPDM activity was induced in the order PCB197≥PCB139>PCB88. In this case, the (–)-enantiomers were more potent inducers of BPDM activities than were the (+)-enantiomers, except for PCB139, in which the (+)-enantiomer was more potent than the (–)-enantiomer.

Furthermore, analysis of uroporphyrin accumulation in cultures treated with δ-aminolevulinic acid revealed that the (+)-enantiomer of PCB139 caused the greatest percentage of URO accumulation, which also correlated with the greatest increase in EROD activity. All other enantiomers caused up to 47% URO accumulation, which did not correlate with an increase in EROD activity. In addition, chiral effects in the induction of drug-metabolising enzymes by atropisomeric PCBs have been reported by Püttmann et al. [32].

In conclusion, much evidence is available that supports the assumption that chirality of atropisomeric PCBs plays an important role in many recognition events associated with enzymatic transformation processes. However, systematic investigations are necessary in order to develop risk assessment concepts that may compete with the TEF concept used for coplanar PCBs. A conceivable experimental approach may include the determination of the EROD activity.

Risk assessment studies of the toxic potential of PCBs suggested that even after having included coplanar and atropisomeric PCBs, a link was still missing. The search for this link drew attention to the PCB metabolites, in particular the PCB methyl sulfones ($MeSO_2$-PCBs). These metabolites are formed via the mercapturic acid pathway, C-S-lyase cleavage of the sulfur–carbon bond in the cysteine moiety, methylation and oxidation of the methyl sulfide as described by Bakke and Gustafsson [33]. The $MeSO_2$-PCBs thus formed are persistent and only slightly less hydrophobic than the parent compounds, which makes them long-lasting contaminants of the biosphere. The separation of $MeSO_2$-PCB enantiomers as well as enantioselective analyses of biota sample extracts are described in Section 3.2.1.2.

From the toxicological point of view several of the 3-$MeSO_2$-PCBs have been shown by Kato and co-workers [34, 35] to induce strongly P-450 cytochrome enzymes such as P-450 2B1, 2B2, 3A2, and 2C6. A recent study on the influence of $MeSO_2$-PCBs on reproduction in mink (*Mustela vison* L.) indicated that a PCB and DDE methyl sulfone mixture may actually increase the litter size in these animals [36]. The mixture also influenced some endocrinological parameters in both dam and offspring. The strong localisation of $MeSO_2$-PCBs to the lungs has been discussed in relation to respiratory distress observed among Yusho patients in Japan [37]. It can thus be concluded that several indications of health effects have been reported, and, consequently, it is relevant to continue studying $MeSO_2$-PCBs, particularly those with specific tissue retentions, in particular placing emphasis on the enantioselectivity of the toxic effects summarised above.

Aimed at the investigation of enantioselective toxic effects induced by chiral $MeSO_2$-PCBs, a Swedish/German cooperative project started in 1997. During the first part of this joint endeavour, highly enantioselective accumulation or formation of $MeSO_2$-PCBs in liver tissues of humans and rats were proved [38]. Meanwhile, a separation of the enantiomers of 3-$MeSO_2$-2,2',3',4',5,6-hexachlorobiphenyl (abbreviation: 3–132), 3-$MeSO_2$-2,2',4',5,5',6-hexachlorobiphenyl (3–149), and 4-$MeSO_2$-2,2',3,4',5',6-hexachlorobiphenyl (4–149), which are highly enriched in liver tissue, was achieved on a preparative basis by enantioselective

HPLC. Thus, sufficient amounts of the separated enantiomers of these three PCB metabolites are now available for extensive investigations of enantioselective toxic effects induced by these metabolites.

4.5
Ecotoxic Effects of Biogenic Chiral Compounds: a Challenge for Enantioselective Chromatography

The assumption often expressed in ecotoxicological reviews that only chemicals released by humans have a hazardous potential in the environment is not correct. Some of the most potent toxins originate from biological sources. For example, snake toxins and venoms are by definition toxic chemicals which usually lead, after oral application, to the death of the prey. A comprehensive review of snake venoms from an evolutionary point of view is given by Kardong [39].

Among the venoms thus far identified, snake venoms consist of very complex protein structures which implies that they are chiral. Whether or not enantiospecific hazardous effects exist is not yet known: due to the complexity of the three-dimensional protein structures the limits and challenges of modern enantioselective detection methods are obvious. Using 'state-of-the art' analytical methods only chemical structures with molecular weight below 10,000 can be investigated routinely. Highly complex three-dimensional protein structures composed of a large number of proteinaceous substructures are difficult to cope with when determining the absolute structures of these toxic compounds. This restriction is one of the major challenges in any future scientific work related to the enantioselective effects of biotoxins.

In the marine environment very potent chiral biotoxins are known, among them tetrodotoxin and saxitoxin (Fig. 4.8). Tedrodotoxin is produced by microorganisms (*Pseudomonas* spp.) and accumulated in puffer fish, fugu fish and other marine organisms via the food chain or by symbiontic mechanisms. Saxitoxin is mainly produced by dinoflagellates (*Gonyaulax catenella, G. tamarensis*), and it is more hazardous than the former biotoxin, since it may be accumulated in filter-feeding shellfish that are cultured for human consumption all over the world. Therefore, incidents are known where people have suffered from severe toxin poisoning after consumption of shellfish, and shellfish farms have been closed down after impact of saxitoxin. Consequently, extensive monitoring programs were initiated and are still going on to control saxitoxin-producing organisms. It is conjectured that saxitoxin can only be produced by the dinoflagellates in a symbiosis with a bacterium. As a result of this highly enantioselective process, only the (+)-enantiomer is found in the marine environment. In mice a LD_{50} value of 5–10 μg/kg has been reported [40]. Thus, saxitoxin is about one order of magnitude more toxic than most man-made nerve gases. It is a very selective Na^+-channel blocker with no effects on K^+- and Ca^{2+}-channels. Nowadays, saxitoxin is one of the most efficient drugs used for the characterisation of Na^+-channel compounds in medical research.

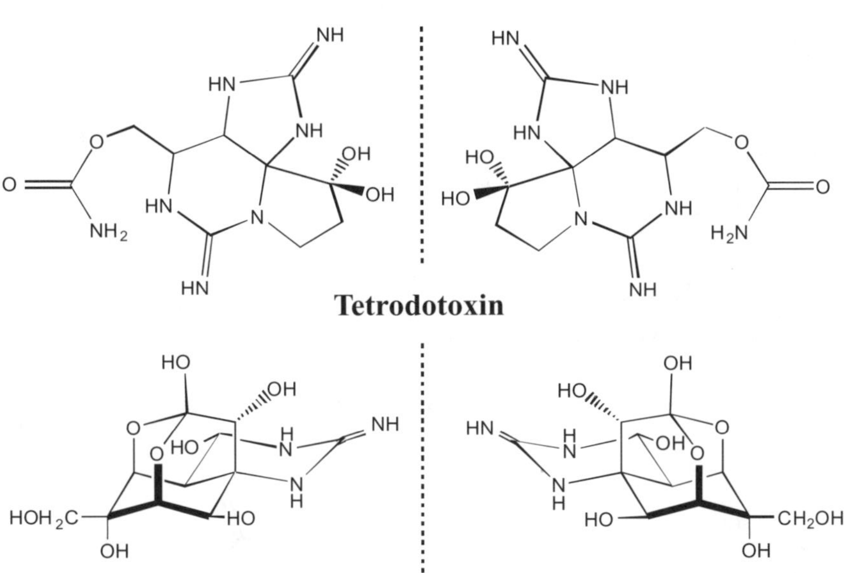

Fig. 4.8. Molecular structures of the marine biotoxins saxitoxin and tetrodotoxin

From an analytical point of view, enantioselective separations of biotoxins are still a challenge because, in general, several asymmetric centres are encountered (see Fig. 4.8). Therefore, several diastereomeric pairs of enantiomers have to be separated. However, toxicological investigations with enantiopure biotoxins would allow deepened insight into the pharmacodynamic and pharmacokinetic properties of the enantiomers of biotoxins.

Potent chiral biotoxins may also be found in limnic ecosystems, in particular if human activities give rise to eutrophication of lakes and rivers caused by the effluent of nutrients. In the course of this process algae blooms often occur leading to a deficiency of oxygen and consequently extinction of life in the respective water system. Another consequence of eutrophication may include a massive development of toxin-producing algae, such as cyanophytes (blue-green algae). In particular, species like *Anabaena flos-aquae* (LYN.) and *Oscillatoria formosa* BO are known to produce strong neurotoxins, e.g., anatoxin-*a* and homoanatoxin-*a* (Fig. 4.9) that are acetylcholinesterase inhibitors [41]. The presence of these neurotoxins can cause a massive reduction in freshwater fish populations and there have even been reports from Scandinavia and the UK [41–43] of the death of cattle and dogs who have consumed the poisoned water. The obvious risk for mammals, including humans, has led to strong efforts to control and monitor the occurrence of toxin-producing cyanophytes in limnic waters. For anatoxin-*a* and homoanatoxin-*a*, LD_{50} values for mice between 200–300 µg/kg clearly demonstrate the high toxic potential of these biotoxins.

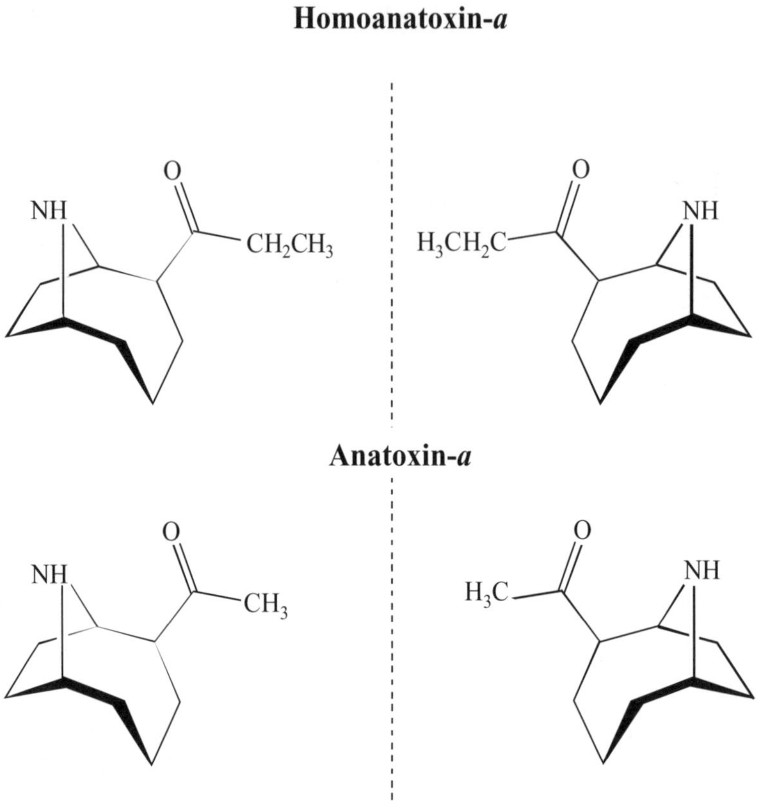

Homoanatoxin-*a*

Anatoxin-*a*

Fig. 4.9. Molecular structures of the chiral limnic biotoxins homoanatoxin-*a* and anatoxin-*a*

Enantioselective analyses of homoanatoxin-*a* were reported by Haugen et al.
[44], who investigated samples which had originated from a toxin-producing
population of the blue-green algae *Oscillatoria formosa* BO from a Norwegian
lake. Clones were cultivated at the Norwegian Institute for Water Research, and
stock cultures were maintained under approved conditions. Isolation of the ho-
moanatoxin-*a* from the algae cultures and its quantitative determination are described [44]. The enantioselective gas chromatographic separation was achieved
with a 12 m glass capillary column coated with a mixture consisting of 20% of
heptakis(2,3,6-tri-*O*-methyl)-β-cyclodextrin dissolved in 85%-methyl-15%-
phenylpolysiloxane (PS086). Racemic synthetic homoanatoxin-*a* was used to
study the enantiomer separation performance of the chiral selector. The hydrochloride salt of (+)-anatoxin-*a* isolated from a NRC-44–1 strain culture of *Anabaena flos-aquae* (LYN.) (Biometric Systems Inc., Minnesota, USA) was used as
an internal standard for the quantification of homoanatoxin. It was added to *Oscillatoria* sample extracts prior to sample clean-up. Quantification of the

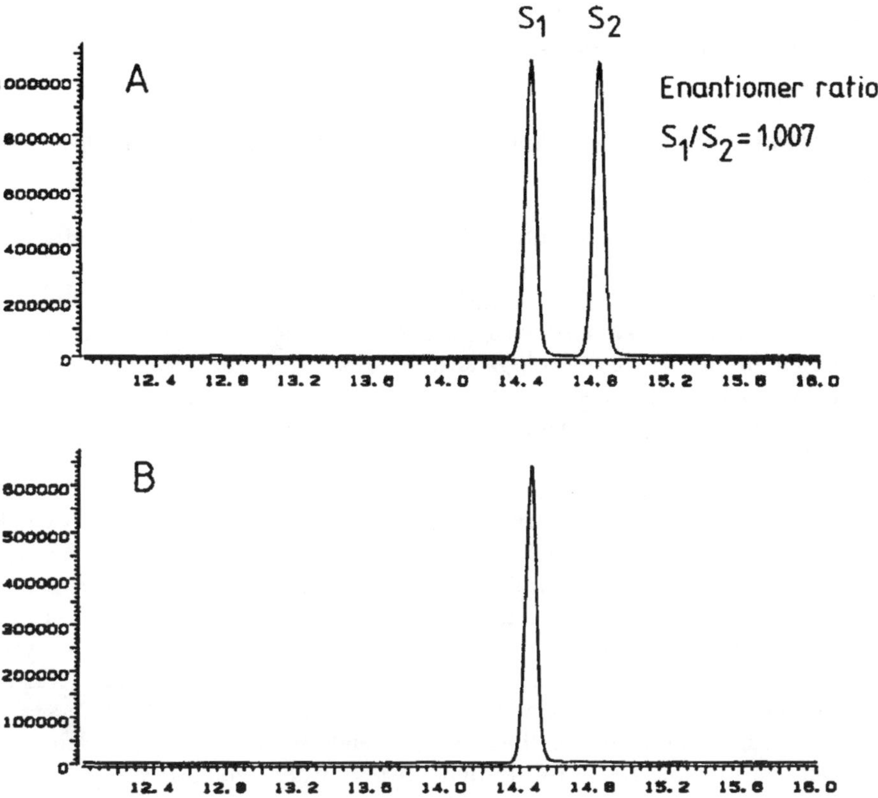

Fig. 4.10A, B. Enantioselective separation of homoanatoxin-*a* obtained by **A** synthesis and **B** from a water extract of a cell culture. Mass *m*/*z* 315 [M-3HF]⁻ was used to monitor the compounds in negative ion chemical ionisation mode. The enantiomeric ratio of the racemate is marked; from [45]

amount of toxin present in a sample was achieved by means of multilevel calibration based on peak area ratios of (+)-anatoxin-*a* and synthetic (±)-homoanatoxin.

As shown in Fig. 4.10, the racemic mixture was completely separated within 15 min. Furthermore, the enantioselective separation of about 100 pg of homoanatoxin-*a*, which was isolated from the toxic cell culture of *Oscillatoria formosa* BO, is also depicted in the figure. Haugen et al. noted that only the first-eluting homoanatoxin-*a* enantiomer appeared in the chromatogram. This result was supported by subsequent investigations of cyanophytes which exclusively produce the (*R*)-(+)-enantiomer of anatoxin-*a*, while the (*S*)-(–)-enantiomer has so far not been found in natural contaminated waters [43, 44]. Furthermore, exclusively enantiopure homoanatoxin-*a* was verified in environmental extracts, whereas racemic homoanatoxin-*a* is obtained by nonenantioselective synthesis

of this compound. However, recently, an asymmetric synthesis has been reported for both (+)- and (–)-anatoxin-*a* [45].

4.6
Differential Mortality of Test Animals Induced by Chiral Environmental Pollutants

A relationship between absolute stereochemistry and biological activity has
been established for hormones and pheromones of juvenile insects since the early 1970s (for Refs. see [46]). These early studies revealed the chiral nature of receptor organs of insects and showed that a specific absolute stereochemistry was
required to induce a specific biological activity. Miyazaki et al., inspired by these
conclusions, raised the question as to whether or not different biological activity
can also be induced by the enantiomers of chiral cyclodiene insecticides and/or
their metabolites such as chlordene, chlordene epoxide, *cis*- and *trans*-chlorda-
ne, and heptachlor *exo*-epoxide [46].

Insecticidal activity of the enantiomers of cyclodiene compounds, which
were prepared by Miyazaki et al. during the first phase of the extensive investigation, was measured on male adults of German cockroach (*Blattella germanica*
L.). A sample of the respective enantiomer and racemate, dissolved in 1.25 µL of
acetone, was applied topically to 10 to 20 insects per dose, and the bioassay was
replicated twice. The toxicity was expressed as percentage mortality at 24 h after
application of the sample, and none of the control insects treated with acetone
only showed mortality under these experimental conditions. The results are
summarised in Table 4.1. (+)-Chlordene, (–)-chlordene epoxide, and (+)-hep-
tachlor *exo*-epoxide, all of which exhibit the same absolute stereochemistry,
showed stronger activity than the corresponding racemates and antipodes.
Among them, a remarkable difference was found between the enantiomers of
chlordene as well as chlordene epoxide. The $LD_{28.6}$ value of (+)-chlordene was
calculated to be 129 µg/g, and the LD_{50} values of (–)-chlordene epoxide and its
racemate to be 76 and 157 µg/g, respectively.

Miyazaki et al. concluded from their results that the insecticidal activity of
these racemates is attributed mainly to the active enantiomer contained in the
racemate, whereas the optical antipodes did not show appreciable toxicity. LD_{50}
values of (+)-, racemic and (–)-heptachlor *exo*-epoxide were calculated as 1.29,
1.82, and 2.98 µg/g, respectively. In the case of heptachlor *exo*-epoxide, the difference between the enantiomers was not so remarkable (about 2.3 times) compared with the above two compounds. In subsequent metabolic studies [47], Miyazaki et al. showed that only the (–)-enantiomer of chlordene epoxide is insecticidal without any bioactivation, whereas the (+)-enantiomer of chlordene becomes toxic as a result of biochemical transformation into the corresponding (–)
-chlordene epoxide.

Slightly modified experimental conditions were used for the investigation of
the insecticidal activity of the enantiomers of heptachlor, 2-chloroheptachlor
and 3-chloroheptachlor: A sample of the respective enantiomer and racemate,

Table 4.1. Insecticidal activity of the enantiomers of chlordene, chlordene epoxide, and heptachlor *exo*-epoxide. The toxicity is expressed as percentage mortality at 24 h after application of the sample; from [46]

Sample	Dose (μg/g)					$LD_{28.6}$ (μg/g)
	300	233	178	126	94.4	
(+)-Chlordene	100%	94.4%	72.2%	33.3%	11.1%	129
Racemic chlordene	28.6%					
(−)-Chlordene	0.0%					

Sample	Dose (μg/g)					LD_{50} (μg/g)
	−	200	133.3	88.9	59.3	
(−)-Chlordene epoxide	−	86.7%	73.3%	66.7%	33.3%	76
Racemic chlordene epoxide	−	66.7%	40.0%	13.3%	0.0%	157
(+)-Chlordene epoxide	−	0.0%				

Sample	Dose (μg/g)					LD_{50} (μg/g)
	6.0	3.0	1.5	0.75	0.375	
(+)-Heptachlor *exo*-epoxide	100%	86.7%	46.7%	33.3%	6.7%	1.29
Racemic heptachlor *exo*-epoxide	93.3%	80.0%	33.3%	13.3%	0.0%	1.82
(−)-Heptachlor *exo*-epoxide	93.3%	46.7%	13.3%	0.0%	0.0%	2.98

dissolved in 0.52 μL of acetone, was applied topically to 10 to 30 male German cockroaches per dose. The bioassay was carried out once, and these results were confirmed by repeated experiments. In the case of (+)-heptachlor, LD_{50} was determined by probit analysis of percentage mortality at 48 h after application of chemicals (for details, see [48]).

The insecticidal activity of the enantiomers and racemates of these cyclodiene derivatives thus determined is summarised in Table 4.2. The (+)-enantiomer of heptachlor was more insecticidal than the (−)-antipode, just as in the case of heptachlor *exo*-epoxide. The weaker activity of (+)-, (−)-, and racemic heptachlor, compared with that of heptachlor *exo*-epoxide, may support the hypothesis that heptachlor becomes insecticidal as a result of bioactivation to heptachlor *exo*-epoxide [48]. Furthermore, it is interesting to note that racemic heptachlor was more insecticidal than either of its enantiomers. Although the difference in toxicity between the racemate and the enantiomers was small, a statistical evaluation showed it to be significant. Miyazaki et al. stress that these data differed entirely from those of other cyclodiene insecticides investigated by their group. In all other cases, the toxicity of the racemate was intermediate between the enantiomers. As a consequence, the authors conjecture that a synergistic action operates between the enantiomers in the toxicity-inducing processes, either by promoting a metabolic activation or by suppressing a detoxification reaction. In order to support their argument, they refer to comparable synergistic effects reported in the literature for an enantiomeric interaction of chiral

Table 4.2. Insecticidal activity of the enantiomers of heptachlor and 2-chloroheptachlor. The toxicity is expressed as percentage mortality at 24 h after application of the sample; from [48]

Sample	Dose (µg/g)						LD_{50}
	18.0	10.8	6.48	3.88	2.32	1.39	(µg/g)
(+)-Heptachlor	93.3%	66.7%	43.3%	0%	0%	0%	3.38
Racemic heptachlor	86.7%	93.3%	60.0%	25.0%	0%	0%	2.64
(–)-Heptachlor	90.0%	46.7%	36.7%	0%	0%	0%	5.32
	200	100	50	25	12.5	–	
(+)-2-Chloroheptachlor	100%	100%	100%	60%	40%	–	20
Racemic 2-chloroheptachlor	100%	80%	50%	10%	10%	–	50
(–)-2-Chloroheptachlor	40%	40%	0%	0%	0%	–	100

pheromone molecules, e.g., in female sex pheromone mimics of redbanded leaf rollers (refs in [48]).

In 2-chloroheptachlor, the (+)-enantiomer with the same absolute stereochemistry as the more toxic (+)-heptachlor exhibited a stronger activity than the (–)-antipode, and that of the racemate was intermediate between them. On the other hand, neither optically active nor racemic 3-chloroheptachlor showed activity at a 200 µg/g dose. The absence of toxicity in the latter case was surprising, taking into account its close structural similarity to 2-chloroheptachlor. Miyazaki et al. ascribe the reason to the different susceptibility of 3-chloroheptachlor for an *in vivo* epoxidation which should be essential for becoming insecticidal.

After these pioneering investigations, Miyazaki et al. expanded their studies to organophosphorus pesticides; the analyses as well as the resolution of which was achieved using enantioselective HPLC [49, 50]. For example, both optically pure enantiomers of methamidophos (*O,S*-dimethyl phosphoramidothioate) and acephate (*O,S*-dimethyl *N*-acetylphosphoramidothioate) were prepared by resolution via diastereomers with L-proline ethyl ester and by HPLC using a chiral adsorbent, respectively [49]. The (*R*)P-(+) enantiomers were more potent to houseflies than the optical antipodes and racemates. In contrast, the (*S*)P-(–)-enantiomers were more toxic within 12 h after application against German cockroaches (*Blattella germanica* L.), although the 24-h LD_{50} values were not significantly different among both enantiomers and racemates of these two organophosphorus compounds.

Stereoselectivity of sulfoxidation of propaphos into propaphos sulfoxide was analysed by enantioselective HPLC using Chiracel OA as chiral stationary phase [50]. Both enantiomers of the sulfoxide were prepared by optical resolution with the same chiral HPLC phase. The (–)-enantiomer was about 3-fold more potent as an inhibitor of German cockroach and bovine erythrocyte acetylcholinesterases compared with the (+)-enantiomer, whereas the (+)-enantiomer was 2-fold more insecticidal to the German cockroach. There was little difference in

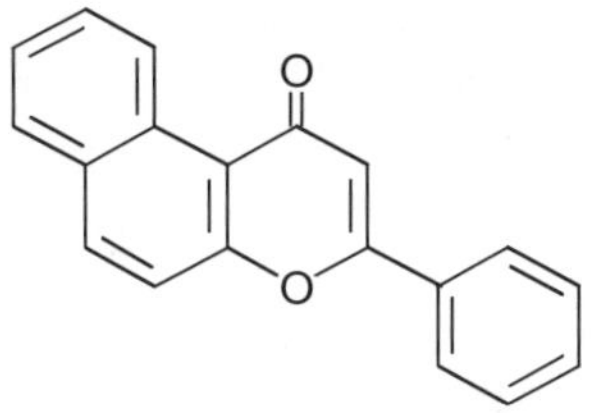

Fig. 4.11. Molecular structure of β-naphthoflavone, used by Pfaffenberger to induce EROD activity in turbot livers during test experiments in laboratory seawater tanks [52]

toxicity to the housefly and the green leaf hopper between the two enantiomers of propaphos sulfoxide.

Pfaffenberger et al. [51] analysed enantiomeric ratios of α-HCH in liver samples from flounders caught in the highly polluted River Elbe estuary (ER= 0.80 ± 0.05) and in the less impacted River Eider estuary (ER=0.89 ± 0.03). These results seemed to support the assumption that a stronger enzymatic activity is induced in the livers of flounders from higher polluted areas. In order to verify this hypothesis, Pfaffenberger carried out a systematic investigation in laboratory seawater tanks using the following experimental approach [52]: the tests were carried out with turbots (*Scophthalmus maximus* L.), which can be well kept in such tanks without significant mortalities for periods of about 100 days. The animals stemmed from the autors' own cultures, grew up under controlled conditions and, as a consequence, exhibited very low concentrations of environmental pollutants.

A suitable parameter for MFO activities in the livers represents the 7-ethoxy-resorufin-O-deethylase (EROD) activity, expressed in "nmol Res/min×mg protein". In contrast to terrestrial animals of higher trophic levels, in fish EROD activity can only be induced by planar molecules such as PAHs, dioxins, dibenzo-furans and coplanar PCBs. In test experiments often a model compound for such planar environmental pollutants is used, for example, β-naphthoflavone (Fig. 4.11), in order to stimulate increased microsomal activity in the liver. Pfaffenberger conjectured that this might in turn lead to increased enantioselective transformation of chiral pollutants. Therefore, five tests were carried out with 25 turbots each weighing about 130 g, which were treated intraperitoneally with the following substances/mixtures:

- *test 1:* α-HCH (30 µg/kg body weight), solved in sunflower oil;
- *test 2:* β-naphthoflavone (β-Nf; 25 mg/kg body weight), solved in sunflower oil;
- *test 3:* a mixture of β-naphthoflavone (25 mg/kg body weight)+α-HCH (6 µg/kg body weight), solved in sunflower oil;
- *test 4:* a mixture of β-naphthoflavone (25 mg/kg body weight)+α-HCH (6 µg/kg body weight)+*cis*-chlordane (6 µg/kg body weight)+*trans*-chlordane (6 µg/kg body weight)+heptachlor (6 µg/kg body weight), solved in sunflower oil; and
- *test 5:* control: pure sunflower oil.

The test period comprised 100 days, where test fish from each series were caught after 12, 21, 28, 42, 54, 77, and 97 days. These fish were sacrificed, and their livers were divided into two parts, one of which was homogenised, extracted, and then subjected to enantioselective analysis, while the second half was used for the determination of EROD activity.

The mortality of the fish encountered in the five test series is summarised in Table 4.3. It is noteworthy that the highest mortality rate was encountered in *test series 3*, with 64% after a period of 15 days. A comparable rate was observed for *test series 4* (56% after 17 days), while for *test series 2* a rate of 40% was found after 32 days. During the complete *test series 1* (α-HCH) and *5* (control=pure sunflower oil) no turbots died in the course of 100 days. These results suggest that β-naphthoflavone is already toxic, but addition of α-HCH or a mixture of α-HCH and the cyclodiene pesticides seems to augment the toxicity. On the other hand, the dose of 30 µg/kg body weight applied in the case of α-HCH did not induce any mortality in the test fish.

The periodic development of the concentrations of α-HCH, *cis-* and *trans-*chlordane, heptachlor, as well as those of the metabolites oxychlordane and heptachlor *exo*-epoxide, in turbot livers of the test animals used in the five test series is summarised in Table 4.4. The background concentration of α-HCH comprised about 4 ng/g wet weight. This value was found in *test series 5* (control) and *2* (β-Nf), which implies that no additional input of α-HCH occurred during the application of sunflower oil and β-naphthoflavone, respectively. Owing to the high mortality rates encountered during *test series 2, 3,* and *4*, concentrations could only be determined until the 21st or 28th day of the experimental period. The most notable result which can be inferred from Table 4.4 is the dramatic increase in α-HCH concentrations during the first 12 days of *test series 1* and *3* (about factor 14). Thereafter, a continuous decrease in α-HCH concentrations

Table 4.3. Mortality of fish as encountered during the five test series, with 25 turbots each; from [52]

Exp. Period (days)	Percentage of dead fish				
	Test 1	Test 2	Test 3	Test 4	Test 5
7	0	4	4	0	0
8	0	8	32	4	0
9	0	16	44	12	0
10	0	24	52	20	0
11	0		60	28	0
12	0			44	0
14	0	32			0
15	0	36	64	48	0
16	0			52	0
17	0			56	0
32	0	40			0

Test 1 α-HCH; *Test 2* β-Nf; *Test 3* β-Nf+α-HCH; *Test 4* β-Nf+α-HCH+*cis-* and *trans-* chlordane+heptachlor; *Test 5* control

was observed in *test series 1*, finally approaching the background level after about 100 days.

The concentrations (in ng/g wet weight) of the cyclodiene pesticides *cis*- and *trans*-chlordane, heptachlor, as well as those of the metabolites oxychlordane and heptachlor *exo*-epoxide, in turbot livers as determined for *test series 4* are summarised in Table 4.4. The background levels of these compounds determined at the start of the test series were below the detection limits. The same result was obtained in the course of the complete experimental period in the *test series 2* (β-Nf; therefore, values not included in Table 4.4) and 5 (control; therefore, values not included in Table 4.4). The highest accumulation was observed for *cis*-chlordane, followed by *trans*-chlordane and heptachlor, with comparable levels. With regard to the transformation products of heptachlor, only the heptachlor *exo*-epoxide was found. Insofar, the same result was obtained as already reported for marine and terrestrial animals of higher trophic levels.

The periodic development of the EROD activity [nmol Res/(min×mg protein)] in turbot livers as determined for the five test series is shown in Table 4.5. A significant increase in the MFO activity was observed for *test series 2* (factor 5), 3 (factor 6.5), and 4 (factor 4), while during *test series 1*(α-HCH only) and 5 (control) no significant effect was encountered. These results imply that, in line with expectations, α-HCH is not capable of inducing increased EROD activity. Thereafter, a decrease in EROD activity was encountered, approaching the original levels after about 28 days.

Table 4.4. Periodic development of the concentrations (in ng/g wet weight) of α-HCH, *cis*- and *trans*-chlordane, heptachlor, as well as of the metabolites oxychlordane and heptachlor *exo*-epoxide, in turbot livers as determined for the five test series; LOD = limit of detection; from [52]

| Days | Test 1 | Test 2 | Test 3 | Test 4 | | | | | | Test 5 |
| | | | | concentrations of: | | | | | | |
	α-HCH	α-HCH	α-HCH	α-HCH	*cis*-chlor-dane	*trans*-chlor-dane	hepta-chlor	oxy-chlor-dane	hepta-chlor *exo*-epoxide	α-HCH
0	4.3	4.3	4.3	4.3	<LOD	<LOD	<LOD	<LOD	<LOD	4.3
11	–	–	–	–	24.7	13.7	14.2	3.3	2.8	–
12	49.0	3.3	58.2	16.6	–	–	–	–	–	4.1
21	42.2	4.0	45.2	21.3	31.0	13.4	12.0	5.3	4.3	–
28	14.8	4.6	53.5	–	–	–	–	–	–	3.6
42	–	–	–	–	–	–	–	–	–	3.8
54	9.2	–	–	–	–	–	–	–	–	2.7
77	9.9	–	–	–	–	–	–	–	–	4.3
97	5.5	–	–	–	–	–	–	–	–	2.1

Test 1 α-HCH; *Test 2* β-Nf; *Test 3* β-Nf+α-HCH; *Test 4* β-Nf+α-HCH+*cis*- and *trans*-chlordane+heptachlor; *Test 5* control

Table 4.5. Periodic development of the EROD activity [nmol Res/(min×mg protein)] in turbot livers as determined for the five test series; from [52]

| Exp. Period | EROD activity | | | | |
(days)	Test 1	Test 2	Test 3	Test 4	Test 5
0	0.037	0.037	0.037	0.037	0.037
12	0.016	0.179	0.240	0.152	0.025
21	0.028	0.026	0.078	0.138	–
28	0.022	0.049	–	0.033	0.013
42	–	0.021	–	–	0.043
97	0.018	–	–	–	0.014

Test 1 α-HCH; *Test 2* β-Nf; *Test 3* β-Nf+α-HCH; *Test 4* β-Nf+α-HCH+*cis*- and *trans*-chlordane+heptachlor; *Test 5* control

In order to verify the hypothesis as to whether an increase in EROD activity correlates with a stronger enantioselective transformation of chiral pollutants, and thus a significant depletion of one enantiomer, the enantiomeric ratios of α-HCH, *cis*- and *trans*-chlordane, heptachlor, as well as of the metabolites oxychlordane and heptachlor *exo*-epoxide, were determined for the turbot liver extracts obtained in the five test series [52]. The stationary phase used in capillary GC consisted of a 1:1 mixture of octakis(3-*O*-butyryl-2,6-di-*O*-*n*-pentyl)-γ-cyclodextrin and OV-1701 (Lipodex E).

In *test series 1, 3,* and *4,* during which α-HCH was injected into the test turbots, a significant and time-dependent depletion of (+)-α-HCH with increasing tendency was observed (Table 4.6). During the first 12 days a faster transformation of the (+)-enantiomer was observed for *test series 4* as compared with *test series 1* and *3.* However, the results summarised in Table 4.6 do not indicate any correlation between the higher EROD activity in *test series 3* and *4* and the ER values for α-HCH. Obviously, the regular level of MFO activities in the liver of turbots is sufficient to allow a continuous metabolisation of α-HCH (see *test series 1*). Additional EROD activity does not give rise to additional transformation of this environmental pollutant.

The enantiomeric ratios of *cis*- and *trans*-chlordane, and of heptachlor, which are also given in Table 4.6, remained at 1.0 throughout the experimental period of 21 days. Presumably, the amount of the respective parent compound that was subject to the transformation in the liver during this period was too small to allow a variation in the ER values. In contrast, the metabolites oxychlordane and heptachlor *exo*-epoxide showed significant deviations from an ER value of 1.0. In both cases, values larger than one were found, which may reflect a preferential formation of the respective (+)-enantiomer by transformation of the parent compound or, alternatively, a preferential transformation of the respective (–)-enantiomer of the two cyclodiene metabolites. The data set presented by Pfaffenberger suggests the simultaneous transformation pathways of both processes: the relatively fast appearance of the metabolites seems to reflect the formation of the (+)-enantiomers. Furthermore, a significant increase in concentrations between the 11th and 21st days supports the hypothesis that the forma-

Table 4.6. Periodic development of the enantiomeric ratios (ER; (+)-/(–)-enantiomer) of α-HCH, *cis*- and *trans*-chlordane, heptachlor, as well as of the metabolites oxychlordane and heptachlor *exo*-epoxide, in turbot livers as determined for the five test series; from [52]

| Days | Test 1 | Test 2 | Test 3 | Test 4 | | | | | | Test 5 |
| | | | | ER of: | | | | | | |
	α-HCH	α-HCH	α-HCH	α-HCH	*cis*-chlor-dane	*trans*-chlor-dane	hepta-chlor	oxy-chlor-dane	hepta-chlor *exo*-epoxide	α-HCH
0	1.00	–	1.00	1.00	1.00	1.00	1.00	–	–	–
11	–	–	–	–	1.00	1.00	1.00	3	2	–
12	0.93	–	0.93	0.81	–	–	–	–	–	–
21	–	–	0.89	0.79	1.00	1.00	1.00	6	2	–
28	0.80	–	0.87	–	–	–	–	–	–	–
54	0.61	–	–	–	–	–	–	–	–	–
77	0.57	–	–	–	–	–	–	–	–	–
97	0.55	–	–	–	–	–	–	–	–	–

Test 1 α-HCH; *Test 2* β-Nf; *Test 3* β-Nf+α-HCH; *Test 4* β-Nf+α-HCH+*cis*- and *trans*- chlordane+heptachlor; *Test 5* control

tion of the metabolites by transformation of the parent compounds is an important process. However, if no other process is active, the ER value should remain constant during the increasing formation of oxychlordane. Therefore, the strong increase in the ER value from 3 to 6 as observed for oxychlordane implies that concurrent transformation of the (–)-oxychlordane must take place as well.

Enantioselective transformation of organic pollutants by liver microsomal cytochromes P-450 was also observed by Shou and Yang [53], who carried out experiments with male Sprague rats (Charles River Breeding Laboratories, Wilmington, MA, USA). The rats, weighing 100–120 g, were treated intraperitoneally with phenobarbital (75 mg/kg body weight, injected in 0.5 mL water) once daily on each of three consecutive days. The rats were sacrificed the day following the last injection of the drug. Liver microsomes were prepared and microsomal protein was determined with bovine serum albumin as protein standard. Incubation of enantiomeric 1-hydroxy-3-methylcholanthrene with these rat liver microsomes resulted in the formation of enantiomeric *trans*-1,2-dihydroxy-3-methylcholanthrene and *cis*-1,2-dihydroxy-3-methylcholanthrene (see Fig. 4.12). The authors investigated these organic pollutants, because 3-methylcholanthrene, a potent mutagen and carcinogen, is metabolised to a complex mixture of nontoxic and mutagenic/tumorigenic products, a considerable part of which consists of the dihydroxy derivatives mentioned above, where the intermediate 1-hydroxy-3-methylcholanthrene appears to play an important role.

Products formed in the incubation experiments were separated using reversed-phase HPLC and further purified by normal-phase HPLC employing a Zorbax silica column. Enantiomeric pairs of 1-hydroxy-3-methylcholanthrene, *cis*-1,2-dihydroxy-3-methylcholanthrene, and *trans*-1,2-dihydroxy-3-methyl-

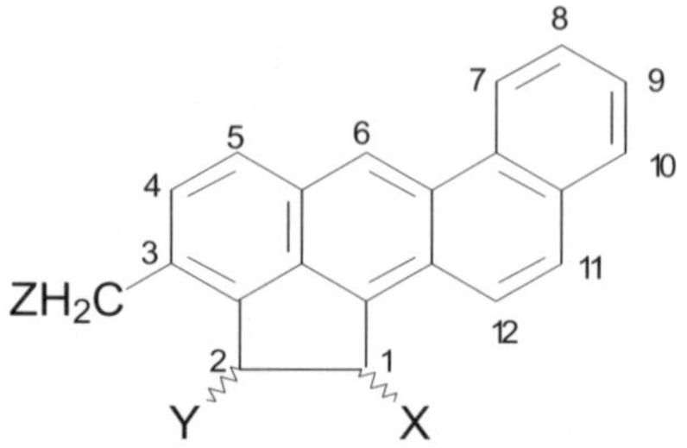

Fig. 4.12. Molecular structures of 3-methylcholanthrene (X=Y=Z=H), 1-hydroxy-3-methylcholanthrene (X=OH; Y=Z=H), *cis*-1,2-dihydroxy-3-methylcholanthrene (X=Y=OH; Z=H), *trans*-1,2-dihydroxy-3-methylcholanthrene (X=Y=OH; Z=H), and 1-hydroxy-3-hydroxymethylcholanthrene (X=Z=OH; Y=H)

cholanthrene were resolved on a CSP column (4.6 mm i.d.×250 mm Rexchrom reversible Pirkle type 1A covalent phenylglycine column; Regis Chemical Co., Morton Grove, IL, USA) packed with spherical particles of 5 μm diameter of γ-aminopropylsilanised silica to which (R)-N-(3,5-dinitrobenzoyl)phenylglycine was covalently bonded.

The absolute configurations of the enantiomeric 1-hydroxy-3-methylcholanthrene and *cis*-1,2-dihydroxy-3-methylcholanthrene were established on the basis of the absolute configuration of an enantiomeric *trans*-1,2-dihydroxy-3-methylcholanthrene sample. Absolute configurations of enantiopure 1-hydroxy-3-hydroxymethylcholanthrene (see Fig. 4.12) were determined by comparing their CD spectra with those of enantiomeric 1-hydroxy-3-methylcholanthrene. This was possible because Shou and Yang were able to separate enantiomeric pairs of synthetic *trans*- and *cis*-1,2-dihydroxy-3-methylcholanthrene by HPLC on a covalently bonded (R)-N-(3,5-dinitrobenzoyl)phenylglycine column. For further experimental details, the reader should refer to [53].

In addition, Shou and Yang could thus verify that the relative amount of three aliphatic hydroxylation products formed by rat liver microsomal metabolism of racemic 1-hydroxy-3-methylcholanthrene was 1-hydroxy-3-hydroxymethylcholanthrene > *cis*-1,2-dihydroxy-3-methylcholanthrene > *trans*-1,2-dihydroxy-3-methylcholanthrene. Enzymatic hydroxylation at C_2 of racemic 1-hydroxy-3-methylcholanthrene was enantioselective towards the (1S)-enantiomer over the (1R)-enantiomer (about 3:1); hydroxylation at the C_3 methyl group was enantioselective toward the (1R)-enantiomer over the (1S)-enantiomer (about 58:48). Rat liver microsomal C_2 hydroxylation of racemic 1-hydroxy-3-methylcholanthrene resulted in *trans*-1,2-dihydroxy-3-methylcholanthrene with a (1S,2S)/(1R,2R) ratio of 63:37 and a *cis*-1,2-dihydroxy-3-methylcholanthrene with a (1S,2R)/(1R,2S) ratio of 12:88, respectively.

Another example of enantioselective transformation processes catalysed by rat liver cytochromes P-450, which were also studied with enantioselective HPLC using the same Pirkle type 1A column containing a 3,5-dinitrobenzoyl-D-phenylglycine chiral stationary phase (Regis Chemical Co., Morton Grove, IL,

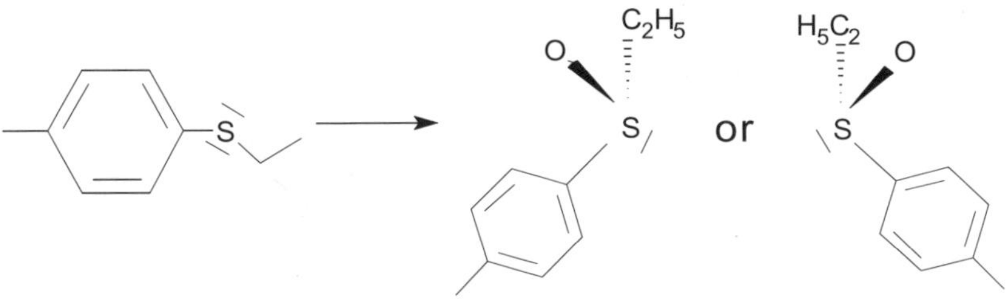

Fig. 4.13. Sulfoxidation of 4-tolylethyl sulfide catalysed by two cytochrome P-450 isozymes, PB-1 and PB-4, purified from phenobarbital-induced rat liver. The products are (R)-4-tolylethyl sulfoxide (*left-hand side*) and (S)-4-tolylethyl sulfoxide (*right-hand side*)

USA), was reported by Waxman et al. [54]. The authors studied the enantioselective sulfoxidation of 4-tolylethyl sulfide applying two cytochrome P-450 isozymes purified from phenobarbital-induced rat liver. Both P-450 isozymes, termed PB-1 and PB-4, when reconstituted with purified rat liver NADPH-cytochrome P-450 reductase and cytochrome b_5, generated 4-tolylethyl sulfoxide (Fig. 4.13), which was predominantly in the (S)-(–)-configuration. In the case of isozyme PB-1, the sulfoxide was $79 \pm 1\%$ S and was formed with a turnover of 41 min^{-1}; with isozyme PB-4, sulfoxide, $84 \pm 1\%$ S, was formed at 31 min^{-1}. In addition, PB-1 catalysed oxygen transfer to the *p*-methyl group of the sulfide substrate to yield the (ethylthio)benzyl alcohol with a turnover of 6.8 min^{-1}, corresponding to a sulfur/carbon oxygenation partition ratio of 6:1. Isozyme PB-4 was about 80-fold less efficient at catalysing this carbon hydroxylation, giving a sulfur/carbon ratio of 375:1. In the absence of cytochrome b_5, turnover numbers were reduced to about 15 and 67% of the above values for PB-1 and PB-4, respectively, with no change in sulfoxide enantioselectivity. Waxman et al. [54] conclude that this fact, and the lack of improvement of enantioselectivity upon inclusion of scavengers for reactive oxygen species, imply that the 79–84% enantioselectivity observed for the sulfoxide product reflects an intrinsic lack of complete stereospecifity in these P-450 catalysed reactions. The enantiomeric composition of 4-tolylethyl sulfoxide generated in rat liver microsomal incubations was shown to reflect the relative contribution of cytochrome P-450 isozymes, which generate the (S)-(–)-enantiomer preferentially, and of the flavin adenine dinucleotide (FAD) containing monooxygenase (EC 1.14.13.8), which catalyses (R)-(+)-sulfoxide formation. Thus, the enantioselectivity of microsome-catalysed sulfoxidation was shown to be modulated by factors which alter the relative participation of these two liver monooxygenases, such as phenobarbital induction, inclusion of inhibitors or activators (metyrapone and *n*-octylamine), and variation in sulfide substrate concentration. Waxman et al. [54] stress that their studies were greatly facilitated by analysis of incubation mixtures on a chiral stationary phase HPLC Pirkle-type column (see above), a method which allows for the rapid, sensitive, and quantitative analysis of enzymatically generated sulfoxide enantiomers.

References

1. Hutt AJ, Hadley MR, Tan SC (1994) Enantiospecific analysis: applications in bioanalysis and metabolism. Eur J Drug Metab Pharmacokin 3:241–251
2. Seiler JP (1995) Chirality – from molecules to organisms. Arch Toxicol Suppl 17:491–498
3. Marzo A (1994) Incoming guidelines on chirality. A challenge for pharmacokinetics in drug development. Arzneimittelforschung 44:791–793
4. Nigrovic V, Diefenbach C, Mellinghoff H (1997) Esters and stereoisomers. Anaesthesist 46:282–285 (German)
5. Aboul-Enein HY, Wainers IW (eds) (1997) The impact of stereochemistry on drug development and use. Wiley, New York, 695 pp
6. Weigel S (1998) Entwicklung einer Methode zur Extraktion organischer Spurenstoffe aus großvolumigen Wasserproben mittels Festphasen. Master thesis (Diplomarbeit), University of Hamburg, 86 pp
7. Annas GJ, Elias S (1999) Thalidomide and the Titanic: reconstructing the technology tragedies of the twentieth century. Am J Public Health 89:98–101
8. Caldwell J (1995) Stereochemical determinants of the nature and consequences of drug metabolism. J Chromatogr A 694:39–48
9. Anonymous (1997) Thalidomide makes a comeback. Stereochem Technol News 4:1–2
10. Domino EF (1986) Opiod-hallucinogen interactions. Pharmacol Biochem Behav 24:401–405
11. Ariëns EJ, van Rensen JJS, Welling W (1988) Stereoselectivity of pesticides. Biological and chemical problems. Elsevier, Amsterdam
12. Ramos Tombo GM, Belluš D (1991) Chiralität und Pflanzenschutz. Angew Chem 103:1219–1241
13. Buser H-P, Francotte E (1996) Stereoselective analysis in crop protection. In: Ahuja S (ed) Chiral separations: applications and technology. ACS Professional Reference Books, Washington, chap 5, pp 93–138
14. Lang D, Criegee D, Grothusen A, Saalfrank RW, Böcker RH (1996) In vitro metabolism of atrazine, terbuthylazine, ametryne, and terbutryne in rats, pigs, and humans. Drug Metab Dispos 24:859–65
15. Bloomquist JR, Adams PM, Soderlund DM (1986) Inhibition of γ-aminobutyric acid stimulated chloride flux in mouse brain vesicles by polychlorocycloalkanes and pyrethroid insecticides. Neurotoxicology 7:11–20
16. Chang RL, Levin VV, Wood AW, Yagi H, Tada M, Vyas KP, Jerina DM, Conney AH (1983) Tumorigenicity of enantiomers of chrysene 1,2-dihydrodiol and of the diastereomeric bay-region chrysene 1,2-diol-3,4-epoxides on mouse skin and in newborn mice. Cancer Res 43:192
17. Möller K, Hühnerfuss H, Wölfle D (1996) Differential effects of the enantiomers of α-hexachlorocyclohexane (α-HCH) on cytotoxicity and growth stimulation in primary rat hepatocytes. Organohalogen Compd 29:357–360
18. Möller K, Hühnerfuss H, Rimkus G (1993) On the diversity of enzymatic degradation pathways of α-hexachlorocyclohexane as determined by chiral gas chromatography. J High Resolut Chromatogr 16:672–673
19. Pfaffenberger B, Hardt I, Hühnerfuss H, König WA, Rimkus G, Glausch A, Schurig V, Hahn J (1994) Enantioselective degradation of α-hexachlorocyclohexane and cyclodiene insecticides in roe-deer liver samples from different regions of Germany. Chemosphere 29:1543–1554
20. Tanabe S, Kannan N, Subramanian A, Watanabe S, Tatsukawa R (1987) Highly toxic coplanar PCBs: occurrence, source, persistency and toxic implications to wildlife and humans. Environ Pollut 47:147–163
21. Safe S (1990) Polychlorinated biphenyls (PCBs), dibenzo-p-dioxins (PCDDs), dibenzofurans (PCDFs), and related compounds: environmental and mechanistic considera-

tions which support the development of toxic equivalency factors. CRC Crit Rev Toxicol 21:51–88

22. Safe S (1992) Development, validation and limitations of toxic equivalency factors. Chemosphere 25:61–64
23. Safe S (1994) Polychlorinated biphenyls (PCBs): environmental impact, biochemical and toxic responses and implications for risk assessment. CRC Crit Rev Toxicol 24:1–63
24. Duinker JC, Schulz DE, Petrick G (1991) Analysis and interpretation of chlorobiphenyls: possibilities and problems. Chemosphere 23:1009–1028
25. Klamer J, Laane RWPM, Marquenie J M (1991) Sources and fate of PCBs in the North Sea: a review of available data. Water Sci Technol 24:77–85
26. Ahlborg UG, Becking GC, Birnbaum LS, Brouwer A, Derks HJGM, Feeley M, Golor G, Hanberg A, Larsen JC, Liem AKD, Safe SH, Schlatter C, Waern F, Younes M, Yrjänheikki E (1994) Toxic equivalency factors for dioxin-like PCBs. Chemosphere 28:1049–1067
27. Falandysz J, Kannan K, Tanabe S, Tatsukawa R (1994) Concentrations, clearance rates and toxic potential of non-*ortho* coplanar PCBs in cod liver oil from the southern Baltic Sea from 1971 to 1989. Mar Pollut Bull 28:259–262
28. Creaser CS, Krokos F, Startin JR (1992) Analytical methods for the determination of non-*ortho*-substituted chlorobiphenyls: a review. Chemosphere 25:1981–2008
29. Rodman LE, Shedlofsky SI, Mannschreck A, Püttmann M, Swim AT, Robertson LW (1991) Differential potency of atropisomers of polychlorinated biphenyls on cytochrome P-450 induction and uroporphyrin accumulation in the chick embryo hepatocyte culture. Biochem Pharmacol 41:915–922
30. De Voogt P, Wells DE, Reutergardh L, Brinkman UATh (1990) Biological activity, determination and occurrence of planar, mono- and di-*ortho* PCBs. Int J Anal Chem 40:1–46
31. Hühnerfuss H, Pfaffenberger B, Gehrcke B, Karbe L, König WA, Landgraff O (1995) Stereochemical effects of PCBs in the marine environment: seasonal variation of coplanar and atropisomeric PCBs in blue mussels (*Mytilus edulis* L.) of the German Bight. Mar Pollut Bull 30:332–340
32. Püttmann M, Mannschreck A, Oesch F, Robertson L (1989) Chiral effects in the induction of drug-metabolizing enzymes using synthetic atropisomers of polychlorinated biphenyls (PCBs). Biochem Pharmacol 38:1345–1352
33. Bakke J, Gustafsson JÅ (1984) Mercapturic acid pathway metabolites of xenobiotics: generation of potentially toxic metabolites during enterohepatic circulation. Trends in Pharmacol Sci 5:517–521
34. Kato Y, Haraguchi K, Kawashima M, Yamada S, Isogai M, Masuda Y, Kimura R (1995) Induction of hepatic microsomal drug-metabolizing enzymes by methylsulphonyl metabolites of polychlorinated biphenyl congeners in rats. Chem-Biol Interact 95:269–278
35. Kato Y, Haraguchi K, Tomiyasu K, Saito H, Isogai M, Masuda Y, Kimura R (1997) Structure-dependent induction of CYP2B1/2 by 3-methylsulfonyl metabolites of polychlorinated biphenyl congeners in rats. Environ Toxicol Pharmacol 3:137–144
36. Lund BO, Örberg J, Bergman Å, Larsson C, Bergman A, Bäcklin BM, Håkansson H, Madej A, Brouwer A, Brunström B (1998) Chronic and reproductive toxicity of a mixture of 15 methylsulfonyl-polychlorinated biphenyls and 3-methylsulfonyl-2,2-bis-(4-chlorophenyl)-1,1-dichloroethene in mink (*Mustela vison*). Environ Toxicol Chem 18:292–298
37. Nakanishi Y, Shigematsu N, Kurita Y, Matsuba K, Kanegae H, Ishimaru S, Kawazoe Y. (1985) Respiratory involvement and immune status in Yusho patients. Environ Health Perspec 59:31–36
38. Ellerichmann T, Bergman Å, Franke S, Hühnerfuss H, Jakobsson E, König WA, Larsson C (1998) Gas chromatographic enantiomer separations of chiral PCB methyl sulfones and identification of selectively retained enantiomers in human liver. Fresenius Envir Bull 7:244–257

39. Kardong KV (1996) Snake toxins and venoms: an evolutionary perspective. Herpetologica 52:36–46

40. Strichartz GR, Hall S, Magnani B, Hong CY, Hishi Y, Debin JA (1995) The potencies of synthetic analogues of saxitoxin and the absolute stereoselectivity of decarbamoyl saxitoxin. Toxicon 33:723–737

41. Carmicheal WW (1981) Freshwater blue-green algae (cyanobacteria) toxins – a review. In: Carmicheal WW (ed) The water environment – algal toxins and health. Plenum Press, New York

42. Skulberg OM, Codd GA, Carmicheal WW (1984) Toxic blue-green algal blooms in Europe: a growing problem. Ambio 13:244–247

43. Haugen J-E, Skulberg OM, Andersen RA, Alexander J, Lilleheil G, Gallagher T, Brough PA (1994) Rapid analysis of cyanophyte neurotoxins: an improved method for quantitative analysis of anatoxin-*a* and homoanatoxin-*a* in the sub-ppb to ppb range. Alg Stud 75:111–121

44. Haugen J-E, Oehme M, Müller MD (1994) Enantiomer-specific analysis of homoanatoxin-*a*, a cyanophyte neurotoxin. In: Codd GA, Jefferies M, Keevil CW, Potter E (eds) Detection methods for cyanobacterial toxins. Proc 1st Symp on detection methods for cyanobacterial (blue-green algal) toxins, 27–29 September, 1993, University of Bath, UK. Royal Society of Chemistry, Cambridge, UK, pp 40–44

45. Koskinen AM, Rapoport H (1985) Synthetic and conformational studies on anatoxin-*a*: a potent acetylcholine agonist. J Med Chem 28:1301–1309

46. Miyazaki A, Hotta T, Marumo S, Sakai M (1978) Synthesis, absolute stereochemistry, and biological activity of optically active cyclodiene insecticides. J Agric Food Chem 26:975–977

47. Miyazaki A, Sakai M, Marumo S (1979) Comparative metabolism of enantiomers of chlordene and chlordene epoxides in German cockroaches, in relation to their remarkably different insecticidal activity. J Agric Food Chem 27:1403–1405

48. Miyazaki A, Sakai M, Marumo S (1980) Synthesis and biological activity of optically active heptachlor, 2-chloroheptachlor, and 3-chloroheptachlor. J Agric Food Chem 27:1310–1311

49. Miyazaki A, Nakamura T, Marumo S (1989) Stereoselectivity in metabolic sulfoxidation of propaphos and biological activity of chiral propaphos sulfoxide. Pestic Biochem Physiol 33:11–15

50. Miyazaki A, Nakamura T, Kawaradani M, Marumo S (1989) Resolution and biological activity of both enantiomers of methamidophos and acephate. J Agric Food Chem 36:835–837

51. Pfaffenberger B, Hühnerfuss H, Kallenborn R, Köhler-Günther A, König WA, Krüner G (1992) Chromatographic separation of the enantiomers of marine pollutants. 6. Comparison of the enantioselective degradation of α-hexachlorocyclohexane in marine biota and water. Chemosphere 25:719–725

52. Pfaffenberger B (1995) Untersuchungen zur enantioselektiven Anreicherung von chiralen organischen Schadstoffen im marinen und terrestrischen Ökosystem. PhD thesis, University of Hamburg, Verlag Shaker, Aachen, Germany, 182 pp

53. Shou M, Yang SK (1990) Enantioselective aliphatic hydroxylations of racemic 1-hydroxy-3-methylcholanthrene by rat liver microsomes. Chirality 2:141–149

54. Waxman D, Light DR, Walsh C (1982) Chiral sulfoxidations catalyzed by rat liver cytochromes P-450. Biochemistry 21:2499–2507

55. Roth HJ, Müller CE, Folkers G (1998) Stereochemie und Arzneistoffe. Wissenschaftl Verlagsgesellschaft mbH, Stuttgart

5 Perspectives of Enantioselective Analyses

5.1
Concern About Chiral Environmental Pollutants and the Legal Implications

In 1990, Brown [1] described the present situation as follows: "Lack of appreciation (but not of knowledge) of the facts that stereoisomers are different compounds and that enantiomers often exert different biological effects has been one factor leading to failure to recognise the significance of chirality in assuring safety, quality and efficacy for medicinal products". Public concern about chiral environmental pollutants can be stated at least since the "Thalidomide/Contergan® case" (see Sect. 4.1). However, adequate regulatory steps have only been taken so far for chiral drugs, while the application of chiral pesticides still needs corresponding regulations both at national and international levels. This lack of regulation seems to be world-wide and results in case-by-case judgments. A survey of the legal implications of chirality is given in several review papers and monographs [1–5]. With regard to the registration of chiral drugs, Witte et al. attempted an overview of the situation in three of the world's most important pharmaceutical areas [4], the most important issues of which are as dealt with in the following sections.

5.1.1
Regulations on Chiral Drugs in the USA

The primary regulatory focus of the Food and Drug Administration (FDA) is claimed to the consideration of both clinical efficacy and consumer safety in making decisions about the allowance of new drugs to be marketed. However, for a long time, the FDA remained blind to questions of stereochemistry. The definition of a drug used by the FDA did not consider the impact of the stereochemical composition of the drug. In 1987, the FDA published a new set of guidelines on the submission of new drug applications (NDAs). For the first time in the history of the FDA, these guidelines mentioned stereochemistry. From that point in time, submissions for a new drug application had to show the molecular structure of the drug and name its chiral centre(s). For enantiomeric ratios different from racemic mixtures, the ratio had to be defined. The proof of the structure had to include the stereochemistry of the compound.

From that date, if a chiral drug was submitted to the FDA as a racemate, studies with the individual enantiomers could be carried out under the existing Notice of Claimed Investigational Exemption for a New Drug (IND) as long as appropriate chemistry, manufacturing, and control data were submitted for the individual enantiomers. In these studies bridging studies could be performed. Yet, the possibility of using bridging data had to be considered on a case-by-case basis by consultation between industry and the authorities. However, for bridging of new drugs which consist of single enantiomers of already marketed racemic drugs, the FDA required a new drug application for the corresponding enantiomer. This meant that a clinical trial had to be carried out under a new IND for that particular enantiomer. Bridging with data of a racemate already on the market was not possible. During the period of these first guidelines, registration of a chiral drug as a racemic mixture or as a single enantiomer was based on a case-by-case basis. The authorities tended to ask for more and more information on the individual enantiomers with regard to their pharmacodynamic and pharmacokinetic profiles.

In 1992, the FDA issued new formal guidelines [3] stating: *The Agency is impressed by the possibility that the use of single enantiomers may be advantageous by permitting better patient control, simplifying dose-response relationships by reducing the extent of interpatient variation in drug response.* The guidelines included the requirement that the chirality of the drug had to be recognised and the stereoisomer producing the activity should be identified. In addition, an enantioselective analytical method and full description of the synthesis, with special attention to the formation of the chiral centre, must be reported, as well as the assignment of the absolute configuration. In addition to the common requirements for all new drug product approvals, in the case of chiral compounds, it is necessary to justify on chemical, pre-clinical and clinical grounds the stereoisomeric form(s) selected for marketing. If a single isomer is chosen, documentation of its synthesis, including formation of the chiral centre, and confirmation of optical stability in the product formulation, on storage and following administration, will normally be required. This means that interconversion to the antipode must be addressed. When a racemate is chosen, justification for its selection in preference to a single isomer should be given. It is also a general requirement to report the pharmacokinetics of individual enantiomers, the occurrence of metabolic chiral inversion *in vivo* (if present), individual pharmacodynamics of each enantiomer (*in vitro* and *in vivo*), and differences in toxicology of the enantiomers. The latter applies when, after establishing the toxicological profile of the racemate, some toxicity not predictable from the pharmacodynamics occurs.

5.1.2
Regulations on Chiral Drugs in the European Community

In 1993, Witte et al. [4] summarised the most important guidelines concerning stereochemistry in the European Community (EC). These were published by the Committee for Proprietary Medicinal Products (CPMP) in Volumes II and III of *The Rules Governing Medicinal Products in the European Community* (for Refs. see [4]). Volume II, *Notice to Applicants for Marketing Authorisations for Medicinal Products for Human Use in the Member States of the EC*, was published in 1989 and contained details concerning the Expert Report. The Expert Report is a specific requirement and is intended to *consist of a critical evaluation of the quality of the product and the investigations carried out in animals and human beings and bring out all the data relevant for evaluation*. It is only in the Expert Report that specific requirements are laid down in relation to chiral drugs in the European Community. It states that where a new active substance contains one or more chiral centres, the stereochemical configuration, i.e., single enantiomer or a mixture of the enantiomers, should be clear. The stereochemical configuration in the final, marketed product should also be clear and be justified. The *Notice to Applicants* also states that, where a mixture of stereoisomers has previously been marketed and it is now proposed to market only one enantiomer, full data on this enantiomer should be provided. Bridging studies with the data of the already marketed racemate are not allowed.

The CPMP guidelines on *Analytical Validation* were adopted in July 1989 and published in the July 1990 Addendum to Volume III (see [4]). The guideline states that "*In all cases, the methods or procedures of analysis must take account of technical and scientific progress and enable the starting material, intermediate and finished product to be checked by means of generally accepted methods*". This guideline does not directly refer to the analysis of single enantiomers in the case of racemic mixtures nor does it mention if, and how, the individual enantiomers ought to be monitored in the body. However, the statement *takes account of technical and scientific progress* and asks for the best and most advanced analytical methods available at the time. In 1993, CPMP issued new formal guidelines. According to Landoni et al. [3], there are many points in common between these guidelines and those published by FDA in 1992.

5.1.3
Regulations on Chiral Drugs in Japan and Other Countries

In 1993, Witte et al. came to the conclusion [4] that, in regard to the situation in Japan, the Japanese regulatory authorities had not yet issued any official statement on the regulation of chiral drugs. However, there was a trend to require more and more detailed information on the differences in pharmacodynamic and pharmacokinetic profiles between the enantiomers. The Japanese authorities recommended an investigation of the absorption, distribution, metabolism and excretion of each of the enantiomers separately and of the racemate, if a racemic mixture was considered for marketing. The enthusiasm for single enantiomers was counteracted by the concern about the cost-benefit balance for the pharmaceutical industry in Japan. Witte et al. expected regulations on chiral drugs to emerge step-by-step in the coming years.

Landoni et al. [3] pointed out that recent discussions between authorities and the pharmaceutical industry, at national and international levels, concerning guidelines for the marketing of chiral compounds, has led to appreciable agreement on the importance to licensing of chirality issues. In this context, they reported that the regulatory authorities of Switzerland, Australia and the Nordic Countries have also promulgated formal guidance, and that Canada was also preparing documentation. The authors raise the question "*Is it predictable that there will be world-wide harmonisation on regulations for chiral compounds in the foreseeable future?*". It has been argued that the *enantiomer versus racemate* debate has made major contributions to the development of safer and more efficacious drug products. It may be that this debate is now largely over, with many companies accepting the need to focus future developments on single stereoisomeric forms. However, we would like to add that *a comparable debate on regulatory guidelines on the approval and application of chiral pesticides is still outstanding and overdue.*

5.2
The Role of Enantioselective Analyses for Model and Mechanistic Studies of Enantioselective Phenomena

Without stereospecific events, prebiotic chemistry would have been restricted to achiral and racemic products. Ariëns claimed that life is predominantly *homochiral* [2, 6]. This latter expression is often used in surface chemistry, in order to describe the exclusive presence of one of the two conceivable enantiomers in an aggregate or in monomolecular surface films or domains of biomembranes. For a deepened discussion on this aspect, the reader is referred to [7] and Section 5.2.1. Single stereoisomers, such as D-enantiomers of carbohydrates, L-amino acids and a few achiral agents, such as purines and pyrimidines, predominate in biochemistry and physiology. By contrast, racemic bioactive compounds are rare. In his review article, Ariëns raises the question "*how did the change from the prebiotic composite chiral* (i.e., racemic) *to the biological homochiral state come about?*". Several models are offered in the literature, and experimental approaches have been suggested aiming at a verification of these models. The crucial test of all these experiments includes the determination of enantiomeric excesses, which have to appear as a result in order to be accepted as a potential explanation for the homochiral preference encountered in life processes [8]. As a consequence, enantioselective analyses have to accompany these experiments. Furthermore, the contradictory postulate has to be solved that the development of enantiomeric excesses requires an asymmetric environment; however, this asymmetric environment itself must be formed in advance, i.e., something like the "chicken-and-egg problem" is encountered [9]. As early as

1923, Bredig et al. denominated this problem, i.e., the enantioselective synthesis from achiral educts without any chiral reagents: *absolute asymmetric synthesis* [10]. Experiments on the template-directed replication of RNA that may simulate the earliest stages of an "RNA world" model for the origin of life are inhibited in the presence of racemic nucleotide monomers, implying that the origin of homochirality must precede the "RNA world" [11]. Although this has been used to argue that homochirality is a prerequisite for the origins of life, it does not exclude the possibility of homochirality being introduced in a "pre-RNA" world such as that based on peptide nucleic acids [11]. We believe that the discussion on absolute asymmetric syntheses may also be relevant to the formation of enantiomeric excesses of environmental pollutants, because some of the mechanisms suggested in the literature would also apply to racemates of pesticides and other prochiral contaminants, provided that one or some of these mechanisms can be verified.

5.2.1
Models for the Prebiotic Formation of Homochirality

Meanwhile, unusual models and experimental approaches have been suggested which may solve the problem of prebiotic formation of homochirality. For example, photochemical reactions may give rise to enantiomeric reactions, a model which at first glance may be surprising considering the results summarised in Section 3.4, clearly indicating nonenantioselective transformation processes. The background to this approach was the observation that in the morning there is a significant "L" circularly polarised component in sunlight, while in the afternoon it reverses to predominantly "D". Since in the afternoon temperatures are appreciably higher than in the morning, a shift in the proportions of the enantiomers, from racemic to homochiral, may occur [12]. The interpretation that this may have formed the basis for biological homochirality and identity in configuration was supported by the syntheses of hepta- and octahelicene with an enantiomeric excess of 7.3% using circularly polarised light [13–15]. In order to maintain homochirality, it is conjectured that natural stereoselective tools in the form of enzymes and receptors operating on a homochiral basis are generated. Ariëns concludes that life would be more complicated if it were based on racemic mixtures of D- and L-amino acids.

Another possibility for potential prebiotic asymmetric syntheses was suggested by Zadel et al. ([16] and literature cited therein) who carried out enantioselective Grignard reactions with prochiral aldehydes in a static magnetic field. The enantioselective excesses determined by ^{1}H- and ^{13}C-NMR using chiral lanthanide shift reagents such as tris(3-heptafluoropropylhydroxymethylene)-D-camphorateuropium(III) or -praseodym(III) attracted the attention of many scientists. The authors claimed to have achieved enantiomeric excesses (*ee*) between 57–98% under these experimental conditions. Reductions of some aryl ketones with lithium aluminum hydride (LAH) proceeded with 11–68% *ee*. Very soon, however, scepticism about these results grew; other research groups tried to repeat the experiments published by Zadel et al., and finally it turned out that the experimental data had been falsified [17, 18]. Furthermore, it should be noted that previous experiments carried out by other research groups in the presence of magnetic fields supplied quite lower enantiomeric excesses, typically below 1% *ee*, and, therefore, it is difficult to accept this phenomenon as a potential route for absolute asymmetric syntheses. We tend to propose that at least asymmetric magnetic fields, the presence of additional magnetic fields etc., must be assumed in order to allow enantiomeric excesses.

Several authors have offered explanations for the predominance of homochiral phenomena in life processes similar to the theoretical background generally accepted for β-decay of neutrons (see [19, 20] and literature cited therein): the parity-violating aspect of the electro-weak interactions; they are assumed to be *principally asymmetric*, because they give rise to an energy difference between enantiomers. However, this conjecture has to be proved yet, as no experimental verification has been published thus far. However, Quack has suggested some experiments, the results of which might verify this hypothesis [19]. But as these are small effects, and would require amplification by factors of about 10^{17} to attain enantiomeric excesses which might account for the homochirality actually found in our world, it remains questionable that weak interactions may have led to such enantiomeric excesses [21]. As an alternative explanation, Podlech [21] raises the question as to whether the origin of homochirality and thus the origin of life is located somewhere in space. As a partial answer he cites results about enantiomeric excesses of about 10% *ee* found by Cronin et al. in meteoritic amino acids using enantioselective cGC/MS analyses [22]. Numerous mechanisms have been proposed to explain this enantiomeric excess. Of these, a hypothesis put forward by Bonner and co-workers [23, 24] seems particularly relevant to the generation of enantiomeric excesses in meteoritic organic compounds. They proposed that large regions in interstellar molecular clouds could be exposed to a flux of circularly polarised light (CPL) of a specific handedness produced as synchrotron radiation by neutron stars. The complex organic mantles of interstellar grains thus could be exposed to ultraviolet CPL, resulting in asymmetric photosynthesis or degradation, that is, asymmetric photolysis of racemic constituents. Bailey et al. suggested that a small enantiomeric excess thus produced by light from solar nebulae could be amplified by some mechanism, which could have led to the origin of homochirality on Earth [11]. A possible mechanism, the majority rule effect, was proposed by Green and Selinger [25]. The prerequisite for this chiral amplification is a stable helix, a conformational state that is common in biological phenomena, which is then subject to the influence of the mixed enantioselective information. The potential role of helical structures in the introduction of chirality in the biosphere was underlined by Schwartz [9] and by Wittung et al. [26]. The latter authors studied DNA analogues with backbones based on peptide linkages, although not of simple α-amino acids. They showed that double-helical complexes are formed between complementary chains of "peptide nucleic acid" (PNA). As the PNA backbone is achiral, the helices can assume equivalent right-handed or left-handed forms (not in DNA where the conformation of deoxyribose plays an important role in favouring a right-handed helix). In the article by Wittung et al. [26], it is shown that the introduction of enantiopure α-amino acid residues at the ends of the PNA chains induces the formation of an enantiomerically uniform set of helices, the sense of each helix being dictated by the chirality of the terminal amino acids. Kinetic measurements of the optical properties observed after mixing complementary, chirally tagged chains, show that two processes are involved: a fast reaction, corresponding to the formation of a racemic set of duplexes, followed by a slower chiral reorganisation of the helices. The authors refer to the reorganisation as *helical seeding*, and suggest that it could have played a major role in the introduction of chirality in the biosphere. But it should be noted that this model requires at least the *chiral seed* and it thus cannot represent the answer to the search for the absolute asymmetric synthesis mentioned above.

Another interesting possibility to induce or amplify asymmetry was reported by Kaupp and Haak [27]: achiral compounds may crystallise in such a way that asymmetric space groups are formed. In some cases, irradiation of such crystals gives rise to the formation of chiral products. It is worth noting that 65 out of 230 space groups fulfill the requirement of being asymmetric, where $P2_12_12_1$ and $P2_1$ belong to the five groups most often encountered for organic compounds. Examples reported by Kaupp and Haak include [2+2]-photodimerisation of achiral compounds with enantiomeric excesses between 0 and 95% *ee*, and even higher values were obtained in some experiments. If this happens without application of any chiral auxiliaries, an *absolute asymmetric synthesis* has taken place. It cannot be excluded that processes of this kind may also be of importance for several classes of organic substances, such as pesticides or other environmental contaminants.

In general, separation of a racemate into its enantiomers by recrystallisation is ruled out, but Ushio et al. [28] succeeded in separating the enantiomers of the antiallergenic drug (±)-2-[4-(3-ethoxy-2-hydroxypropoxy)phenylcarbamoyl-ethyl]dimethylsulfonium-*p*-sulfonate by repeated recrystallisation. The authors cannot explain this phenomenon, they only offer a tentative explanation referring to "the unusual polymorphism" of the crystals of this compound. However, this conjecture has yet to be verified.

Hanein et al. [29] showed that interactions during cell adhesion to external surfaces may reach the level of discrimination of molecular chirality. Cultured epithelial cells interact differently with the {011} faces of the (*R,R*) and (*S,S*) calcium tartrate tetrahydrate crystals. In a modified version of the classical Pasteur experiment, the enantiomorphous crystals were sorted out from a 1:1 mixture by the selective adhesion of cells to the (*R,R*) crystals. This stereospecifity results from molecular recognition between chiral components on the cell surface and the structured crystal surface. The authors conjecture that crystals may allow experimental differentiation between distinct stages in cell substrate contacts, providing mechanistic information not readily attainable on conventional heterogeneous surfaces.

In conclusion, models for the prebiotic formation of homochirality thus far discussed in the literature appear to suggest that there are several mechanisms which could have led to enantiomeric excesses of chiral organic molecules. However, the central problem related to the amplification of the respective *ee*, finally leading to homochiral clusters, is still unresolved. The key step for the formation of homochiral clusters has to include a spontaneous segregation of a racemic mixture. This was actually observed for *two-dimensional systems*, i.e., in Langmuir monolayers consisting of chiral surface-active compounds [7, 30–39]. Systematic investigations by Stine and co-workers [35–37] and by Hühnerfuss and co-workers [7, 30–34] have shown that some important conditions have to be fulfilled in order to allow chiral discrimination of this kind (sometimes referred to as *chiral symmetry breaking*): homochiral interactions (*R/R* or *S/S*) must be preferable over heterochiral interactions (*R/S*). Meanwhile several examples of *N*-acylamino acids have been shown to fulfill this requirement by applying fluorescence microscopy [35–37] or infrared reflection-absorption spectroscopy (IRRAS) [7, 30–34]. All amino acids, the derivatives of which were included in the studies by Hühnerfuss, Stine and their co-workers, belong to the group of proteinogenous amino acids and thus are potential components of biomembranes. Furthermore, these proteins may form helices. Accordingly, if we tentatively reduce the problem of homochiral cluster formation to a monolayer (two-dimensional) or bilayer (membrane) problem, several observations can be inferred from the literature that may supply at least partial answers: Spontaneous segregations of racemic mixtures of proteinogenous amino acid derivatives were reported by Hühnerfuss, Stine and co-workers [7, 30–37]; furthermore, we can evoke the proposals of Green and Selinger [25] (*prerequisite for chiral amplification is a stable helix*) and by Hanein et al. [29] (*cell adhesion as a step towards chiral discrimination*). Presently, Hoffmann (personal communication) is carrying out systematic investigations with chiral surface-active compounds aimed at answering the question: which intermolecular interactions may be the driving forces that give rise to the spontaneous segregation in racemic monolayers. In summary, these two-dimensional mechanisms bear the potential for a spontaneous segregation of racemic mixtures of chiral organic compounds and thus one conceivable possibility for the development of homochirality in our world.

5.2.2
Innovative New Enantioselective Detectors

Achieving chiral discrimination with gas sensors is fundamentally more difficult than with gas chromatography, because the former uses only one "theoretical plate" (one absorption/desorption step), whereas cGC discrimination typically results from the cumulative effect of thousands of successive absorption/desorption equilibria. Gas chromatography, however, is an off-line procedure, whereas gas sensors offer the advantage of fast and continuous monitoring. Thus Bodenhöfer et al. investigated the ability of sensor arrays to discriminate optical enantiomers [40]. They described two different sensor systems that are capable of recognising different enantiomers and of qualitatively monitoring the enantiomeric composition of amino acid derivatives and lactates in the gas phase. One sensor detects changes in mass, due to binding of the compound being analysed, by thickness shear-mode resonance (TSMR); the other detects changes in the thickness of a surface layer by reflectometric interference spectroscopy (RIFS; for refs reporting both methods, see [40]).

The authors point out that their sensors apply both enantiomers of a chiral polymeric receptor simultaneously, while natural odour receptors are proteins that use only one enantiomeric form (the L-form). Their chiral model receptor consisted of octyl-Chirasil-Val®, which was derived from the well-known chiral chromatographic stationary phase Chirasil-Val®. It contains chiral peptide residues for enantiomer recognition and nonchiral lipophilic side chains. Solutions of these polymers were sprayed onto quartz plates (by airbrush) or spin-cast on the optical devices (layer thickness 100 to 300 nm). In addition to the two sets of chiral sensors, reference devices coated with the nonenantioselective polymer poly(dimethylsiloxane) (SE-30) were included in the sensor arrays to recognise artifacts caused by fluctuating gas-phase concentrations or contaminations of the analytes. Both receptor enantiomers are simultaneously exposed to the analyte enantiomer, so a higher and a lower sensor signal are observed at the same time. In switching from one enantiomer of the analyte to the other, a cross-wise inversion of the signals provided by the two chiral polymers is expected. Artifacts or problems in vaporisation, which might lead to chiral discrimination even by nonchiral coatings, can be definitely excluded by this way. Such sensors may prove useful in controlling the enantiomeric purity of anesthetics and drugs in pharmacology, or in monitoring the quality of, for example, lactates, which are synthesised in industrial quantities.

References

1. Brown JR (1990) Drug chirality. Impact on pharmaceutical regulation. Legal Studies and Services Limited, Healthcare & Regulatory Affairs Division, The London Press Centre, London, UK
2. Ariëns EJ (1993) Nonchiral, homochiral and composite chiral drugs. Trends Pharmacol Sci 14:68–75
3. Landoni MF, Soraci AL, Delatour P, Lees P (1997) Enantioselective behaviour of drugs used in domestic animals: a review. J Vet Pharmacol Therap 20:1–16
4. Witte DT, Kees E, Franke J-P, De Zeeuw RA (1993) Development and registration of chiral drugs. Pharmacy World & Science 15:10–11
5. Hutt AJ (1991) Drug chirality: impact on pharmaceutical regulation. Chirality 3:161–164
6. Ariëns E (1986) Stereochemistry: a source of problems in medicinal chemistry. Med Res Rev 6:451–466
7. Hoffmann F, Hühnerfuss H, Stine KJ (1998) Temperature dependence of chiral discrimination in monolayers of N-acylamino acids as inferred from Π/A measurements and infrared reflection-absorption spectroscopy. Langmuir 14:4525–4534
8. Krempaský J, Krejèíová E (1993) On the origin of "pure" chirality of amino acids and saccharides at the prebiotical stage. Gen Pysiol Biophys 12:85–91
9. Schwartz AW (1994) The origin of macromolecular chirality. Curr Biol 4:758–760
10. Bredig B, Mangold P, Williams TG (1923) Über "absolute" asymmetrische Synthese. Angew Chem 36:456–458

11. Bailey J, Chrysostomou A, Hough JH, Gledhill TM, McCall A, Clark S, Ménard F, Tamura M (1998) Circular polarization in star-formation regions: implications for biomolecular homochirality. Science 281:672–674
12. Deutsch DH (1991) A mechanism for molecular asymmetry. J Mol Evol 33:295–296
13. Bernstein WJ, Calvin M, Buchardt O (1972) Absolute asymmetric synthesis. I. On the mechanism of the photochemical synthesis of nonracemic helicenes with circularly polarized light. Wavelength dependence of the optical yield of octahelicene. J Am Chem Soc 94:494–498
14. Bernstein WJ, Calvin M, Buchardt O (1973) Absolute asymmetric synthesis. III. Hindered rotation about aryl-ethylene bonds in the excited states of diaryl ethylenes. Structural effects on the asymmetric synthesis of 2- and 4-substituted hexahelicenes. J Am Chem Soc 95:527–532
15. Kagan H, Moradpour A, Nicoud JF, Balavoine G, Martin RH, Cosyn JP (1971) Photochemistry with circularly polarised light. II. Asymmetric synthesis of octa- and nonahelicene. Tetrahedron Lett 1971:2479–2482
16. Zadel G, Eisenbraun C, Wolff G-J, Breitmaier E (1994) Enantioselektive Reaktionen im statischen Magnetfeld. Angew Chem 106:460–463; Angew Chem Int Ed Engl 33:454
17. Feringa BL, Kellogg RM, Hulst R, Zondervan C, Kruizinga WH (1994) Attempts to carry out enantioselective reactions in a static magnetic field. Angew Chem 106:1526–1527
18. Kaupp G, Marquardt T (1994) Absolute asymmetrische Synthese allein durch ein statisches Magnetfeld? Angew Chem 106:1527–1529
19. Quack M (1989) Struktur und Dynamik chiraler Moleküle. Angew Chem 101:588–604
20. Van House J, Rich A, Zitzewitz PW (1984) Beta decay and the origin of biological chirality: new experimental results. Origin of Life 14:413–420
21. Podlech J (1999) Neue Einblicke in den Ursprung der Homochiralität biologisch relevanter Moleküle – Grundstoffe des Lebens aus dem All? Angew Chem 111:501–502; Angew Chem Int Ed 38:477–478
22. Cronin JR, Pizzarello S (1997) Enantiomeric excesses in meteoritic amino acids. Science 275:951–955
23. Flores JJ, Bonner WA, Massey GA (1977) Asymmetric photolysis of (RS)-leucine with circularly polarised ultraviolet light. J Am Chem Soc 99:3622–3625
24. Rubenstein E, Bonner WA, Noyes HP, Brown GS (1983) Supernovae and life. Nature 306:118
25. Green MM, Selinger JV (1998) Cosmic chirality. Science 282:880–881
26. Wittung P, Nielson PE, Buchardt O, Edholm M, Norden B (1994) DNA-like double helix formed by peptide nucleic acid. Nature:368:561–563
27. Kaupp G, Haak M (1993) Absolute Synthese durch Belichtung chiraler Kristalle. Angew Chem 105:727–728
28. Ushio T, Tamura R, Takahashi H, Azuma N, Yamamoto K (1996) Ungewöhnliche Phänomene bei der Racematspaltung durch Umkristallisation einer racemischen Verbindung. Angew Chem 108:2544–2546
29. Hanein D, Geiger B, Addadi L (1994) Differential adhesion of cells to enantiomorphous crystal surfaces. Science 263:1413–1416
30. Neumann V, Gericke A, Hühnerfuss H (1995) Comparison of enantiomeric and racemic monolayers of 2-hydroxyhexadecanoic acid by external infrared reflection-absorption spectroscopy. Langmuir 11:2206–2212
31. Hühnerfuss H, Neumann V, Stine KJ (1996) The role of hydrogen bond and metal complex formation for chiral discrimination in amino acid monolayers studied by infrared reflection-absorption spectroscopy. Langmuir 12:2561–2569
32. Hühnerfuss H, Gericke A, Neumann V, Stine KJ (1996) The determination of the molecular order of chiral monolayers at the air/water interface by infrared reflection-absorption spectroscopy "IRRAS" – a bridge between physico- and biochemistry. Thin Solid Films 284:694–697

33. Hoffmann F, Hühnerfuss H, Stine KJ (1998) Temperature dependence of chiral discrimination in monolayers of N-acyl amino acids as inferred from Π/A-measurements and infrared reflection-absorption spectroscopy. Langmuir 14:4525–4534

34. Gericke A, Hühnerfuss H (1994) Infrared spectroscopic comparison of enantiomeric and racemic N-octadecanoylserine methyl ester monolayers at the air/water interface. Langmuir 10:3782–3786

35. Stine KJ, Uang JY-J, Dingman SD (1993) Comparison of enantiomeric and racemic monolayers of N-stearoylserine methyl ester by fluorescence microscopy. Langmuir 9:2112–2118

36. Stine KJ, Whitt SA, Uang JY-J (1994) Fluorescence microscopy study of Langmuir monolayers of racemic and enantiomeric N-stearoyltyrosine. Chem Phys Lipids 69:41–50

37. Parazak DP, Uang JY-J, Turner B, Stine KJ (1994) Fluorescence microscopy study of chiral discrimination in Langmuir monolayers of N-acylvaline and -alanine amphiphiles. Langmuir 10:3787-

38. Bringezu F, Brezesinski G, Nuhn P, Möhwald H (1996) Chiral discrimination in a monolayer of a triple-chain phosphatidylcholine. Biophys J 70:1789–1795

39. Groves JT, McConnell HM (1996) Chiral discrimination in two dimensions. Biophys J 70:1573–1574

40. Bodenhöfer K, Hierlemann, Seemann J, Gauglitz G, Koppenhoefer B, Göpel W (1997) Chiral discrimination using piezoelectric and optical gas sensors. Nature 387:577–580

Printing: Mercedes-Druck, Berlin
Binding: Buchbinderei Lüderitz & Bauer, Berlin